Springer Theses

Recognizing Outstanding Ph.D. Research

Aims and Scope

The series "Springer Theses" brings together a selection of the very best Ph.D. theses from around the world and across the physical sciences. Nominated and endorsed by two recognized specialists, each published volume has been selected for its scientific excellence and the high impact of its contents for the pertinent field of research. For greater accessibility to non-specialists, the published versions include an extended introduction, as well as a foreword by the student's supervisor explaining the special relevance of the work for the field. As a whole, the series will provide a valuable resource both for newcomers to the research fields described, and for other scientists seeking detailed background information on special questions. Finally, it provides an accredited documentation of the valuable contributions made by today's younger generation of scientists.

Theses are accepted into the series by invited nomination only and must fulfill all of the following criteria

- They must be written in good English.
- The topic should fall within the confines of Chemistry, Physics, Earth Sciences, Engineering and related interdisciplinary fields such as Materials, Nanoscience, Chemical Engineering, Complex Systems and Biophysics.
- The work reported in the thesis must represent a significant scientific advance.
- If the thesis includes previously published material, permission to reproduce this must be gained from the respective copyright holder.
- They must have been examined and passed during the 12 months prior to nomination.
- Each thesis should include a foreword by the supervisor outlining the significance of its content.
- The theses should have a clearly defined structure including an introduction accessible to scientists not expert in that particular field.

More information about this series at http://www.springer.com/series/8790

Kara Marie Lynch

Laser Assisted Nuclear Decay Spectroscopy

A New Method for Studying Neutron-Deficient Francium Isotopes

Doctoral Thesis accepted by
the University of Manchester, UK

Author
Dr. Kara Marie Lynch
School of Physics and Astronomy
University of Manchester
Manchester
UK

Supervisor
Dr. Kieran Flanagan
School of Physics and Astronomy
University of Manchester
Manchester
UK

ISSN 2190-5053
ISSN 2190-5061 (electronic)
ISBN 978-3-319-36138-3
ISBN 978-3-319-07112-1 (eBook)
DOI 10.1007/978-3-319-07112-1

Springer Cham Heidelberg New York Dordrecht London

Softcover reprint of the hardcover 1st edition 2015

Printed on acid-free paper

Springer is part of Springer Science+Business Media (www.springer.com)

Supervisor's Foreword

Even though more than 100 years has now elapsed since Rutherford discovered the nucleus there remain many unanswered questions associated with the force that binds protons and neutrons together. This directly impacts our understanding of astrophysical processes and the properties of the early universe. This compelling area of research requires both theoretical and experimental efforts. A key component of this work is the measurements of nuclear observables that do not rely on prior assumptions of a particular theoretical model. The interaction between atomic electrons and the nucleus is sensitive to the size, shape, and magnetization of the nucleus. Of the many methods that can be used to measure these nuclear observables, laser spectroscopy is particularly well suited to studying unstable nuclei.

It has now been nearly 40 years since the first laser spectroscopy experiments were carried out on unstable isotopes. During the intervening years many techniques have been developed to measure the nuclear moments, charge radii, and spins of unstable nuclei. Many of these variations on the original concept have been tailored to suit the constraints imposed by a particular element. Two methods stand out in terms of their wide application, resolution, and sensitivity: collinear and in-source laser spectroscopy. Collinear methods offer high resolution while in-source methods offer the highest sensitivity. In 1982 a method for combining the resolution of the collinear method with the sensitivity of the in-source method was proposed. At that time, the advantages of the collinear resonance ionization spectroscopy (CRIS) technique were offset with significant technological challenges. A critical problem was the duty cycle loss associated with using pulsed lasers to probe continuous beams. The introduction of ion-beam cooling and bunching techniques in the last decade resolved this problem and resurrected the collinear resonance ionization concept.

Kara Lynch joined this endeavour in 2010 and focussed her Ph.D. on a fascinating region of the nuclear chart. The neutron-deficient isotopes of francium sit at a pivotal point where a spherical shell model description evolves into a deformed liquid drop. This region is characterized by a softening of the nuclear potential with decreasing neutron number and yet maintaining a spherical shape and almost pure wave functions. As the proton-drip line is approached, it is expected that there will be an abrupt change in the ground-state structure from a spherical to highly deformed shape as the occupation of intruder states becomes energetically favourable. Kara simultaneously applied the techniques of laser spectroscopy and decay spectroscopy to study the $^{202-206}$Fr. These measurements confirm for the first time the existence of the 10^- isomeric states in ^{204,206}Fr and allow analysis of the composition of the nuclear wave function. Selecting the isomers in ^{204}Fr using the CRIS technique, Kara was able to perform decay spectroscopy on just the excited state. This cleaning method provides the route to measuring unambiguously the branching ratio and half-lives of states with overlapping alpha-decay energies. In order to study these exotic species, Kara constructed a bespoke decay spectroscopy station that is compatible with the UHV requirements of the experiment. The experiments reported in Kara Lynch's thesis represent the first successful demonstration of the CRIS technique with a total experimental efficiency of greater than 1 %. By measuring the hyperfine structure of ^{202}Fr, with a production rate of 100 atoms/s, Kara demonstrated that the method is more than two orders of magnitude more sensitive than existing techniques. Kara's research has made a large impact within the community and she has been awarded the prestigious IOP Nuclear Group Prize.

Manchester, June 2014 Dr. Kieran Flanagan

Abstract

Radioactive decay studies of rare isotopes produced at radioactive ion beam facilities have often been hindered by the presence of isobaric and isomeric contamination. The Collinear Resonance Ionization Spectroscopy (CRIS) experiment at ISOLDE, CERN uses laser radiation to stepwise excite and ionize an atomic beam in a particular isomeric state. Deflection of this selectively ionized beam of exotic nuclei, from the remaining neutral contaminants, allows ultra-sensitive detection of rare isotopes and nuclear structure measurements in background-free conditions.

This thesis outlines the work undertaken in the development of the novel technique of laser-assisted nuclear decay spectroscopy. The isomeric ion beam is selected using an atomic resonance of its hyperfine structure, where it is deflected to a decay spectroscopy station. This consists of a rotating wheel implantation system for alpha-decay spectroscopy, and up to three germanium detectors around the implantation site for gamma-ray detection.

Laser spectroscopy provides a measurement of the spin, moments, and change in mean-square charge radii of the ground and isomeric states in the parent nucleus. Complementary information on the level structure of the daughter nucleus comes from the decay spectroscopy, providing further information on the isotope under investigation.

The new techniques of collinear resonance ionization spectroscopy and laser-assisted nuclear decay spectroscopy have been developed and optimized in the experimental campaign studying the neutron-deficient francium isotopes. In this thesis, the hyperfine structure studies of $^{202-207,211}$Fr are presented, alongside the radioactive decay studies of ^{202,204,218}Fr and their isomers.

Acknowledgments

To Kieran, for the enormous support and encouragement he has given me throughout my Ph.D. I couldn't have hoped for a better supervisor, nor a more inspirational one.
To Thomas, for being a constant source of nuclear-physics knowledge, patience, and gossip!
To the CRIS collaboration, for making it all possible. I consider myself very lucky to work with such a talented group of individuals.
To Jon Billowes and Gerda Neyens, for the depth of knowledge and understanding they have brought to my work.
To Yorick Blumenfeld and Maria Garcia Borge, for the wonderful opportunity their support and kindness has given me.
To the on-site team, for making ISOLDE such an inspiring place to work.
To Tom, for all the laughs we shared during our Ph.Ds.
To Andy Smith, for his invaluable help with the DSS design.
To the core, for allowing me to be a valence particle.
To Sarah, for the life-long friendship that started when we were summer students.
To my parents, for their endless love and support. It means the world to me.
To Espen, for making me smile everyday.
To all my friends, family, and colleagues who have made these last 3 years so special…

Thank you! Kara.

Contents

Chapter 1
Introduction

Many experiments in nuclear physics are unable to study rare isotopes due to the presence of large isobaric contamination. Of particular interest are the low-lying isomers of less than 40 keV that cannot easily be distinguished from their ground state. Mass measurements of ^{80}Ga [1] did not observe the long-lived isomer that was subsequently discovered with collinear laser spectroscopy [2]. Gamma-ray spectroscopy provides higher resolution, but cannot distinguish between ground and isomeric states that are too long lived (e.g. more than several mins). Difficulty also arises for states with similar half-lives that radioactively decay via the same mechanism. In the case of ^{80}Ga, the two long-lived states have similar half-lives ($\sim$1 s) and both β-decay to excited states of ^{80}Ge. From the information provided by collinear laser spectroscopy, high-resolution gamma-ray spectroscopy has since be able to determine their spin and parity [3]. In some cases, the resolution of gamma-ray spectroscopy may be insufficient to distinguish the states, for example in ^{73}Ga where the isomeric $3/2^-$ state is near-degenerate with the 1/2 ground state (within 1 keV of each other) [4].

Initial decay spectroscopy of ^{202}Fr and ^{204}Fr could not differentiate the alpha decay of the isomeric states from those of the ground state [5]. However, later measurements [6] revealed the low-lying structure of these isotopes, but the tentative spin-parity assignments of these isomers are based on feeding patterns in β+/EC decay and alpha-decay systematics. Recent measurements of the neutron-deficient francium isotopes have been able to determine the alpha-decay properties of the ground and isomeric states more precisely [7–9], but their exact nature is still unknown.

Isomer identification has already been achieved with in-source laser spectroscopy [10], for example ^{68,70}Cu [11]. Following this selection, secondary experiments such as Coulomb excitation [12] and mass measurements [13] have been performed on these isomeric beams. However, these experiments suffered from the isobaric contamination, as well as significant ground-state contamination due to the Doppler broadening of the hyperfine resonances of each isomer [14].

The status of radioactive ion beam experiments is such that high-resolution laser spectroscopy measurements cannot, in general, currently be performed on short-lived

K.M. Lynch, *Laser Assisted Nuclear Decay Spectroscopy*,
Springer Theses, DOI: 10.1007/978-3-319-07112-1_1

isotopes with yields less than 10^2 atoms per second. The Collinear Resonance Ionization Spectroscopy (CRIS) technique [15], located at the ISOLDE facility (CERN, Switzerland), aims to perform hyperfine-structure measurements on these rare isotopes. Using a combination of atomic and nuclear physics methods, the CRIS experiment uses lasers to probe the hyperfine structure of atoms and to measure nuclear observables. The hyperfine structure can be thought of as an atomic fingerprint: unique for every nuclear state. This provides the ability to separate different isotopes (and isomers) and study very exotic nuclei. The CRIS technique provides a combination of high-detection efficiency, high resolution and ultra-low background, pushing the limits of laser spectroscopy to perform measurements on isotopes at the edges of stability, measuring isotopes with yields below 1 atom per second. Studying the rare neutron-deficient francium isotopes provides the unique opportunity to answer questions about the nuclear structure in this region of the nuclear chart: the shape of the nuclei and their quantum configuration. In addition, selection of the ground or isomeric state present in the radioactive beam, from an atomic resonance of its hyperfine structure, allows decay spectroscopy on pure-state beams to be performed. The CRIS experiment has developed a novel technique of laser assisted nuclear decay spectroscopy, allowing the study of the neutron-deficient francium isotopes and their isomers.

1.1 Physics Motivation for Studying Francium

Francium (Z = 87), the heaviest alkali metal, has a single electron outside a closed atomic shell. Coupled with a large nuclear charge, the francium isotopes are ideal candidates to test our knowledge of the electronic and nuclear structure of atoms.

An atomic test of the standard model, such as the investigation into parity nonconservation (PNC) [16–18], can be performed with francium. Interaction of the electron's wavefunction with the nucleus causes the weak force to destroy the electron's pure-parity state. This results in electron transitions being possible between two energy levels with the same orbital angular momentum. To measure the probability of this parity nonconservation, a detailed knowledge of the atomic structure is needed [19]. The most accurate PNC calculations have been performed on caesium, francium's lighter alkali neighbour. With experiment and theory agreeing within 1%, francium represents a heavier analogue of caesium, with its simple electronic structure, to test the larger PNC effect due to its larger nuclear charge [20].

The time evolution of fundamental constants [21, 22], predicted by unified theories such as superstring theory and M theory, can also be tested with the francium atom. By comparing quasar absorption spectra with spectra measured in the laboratory, it is possible to probe the variations in the fine-structure constant over cosmological time-scales. This is achieved by comparing the observed transitional wavelength of alkali doublets [23, 24].

With no stable isotope of francium present in nature, and the longest-lived isotope of ^{223}Fr having a half-life of 22 min, the energy levels of francium are the least

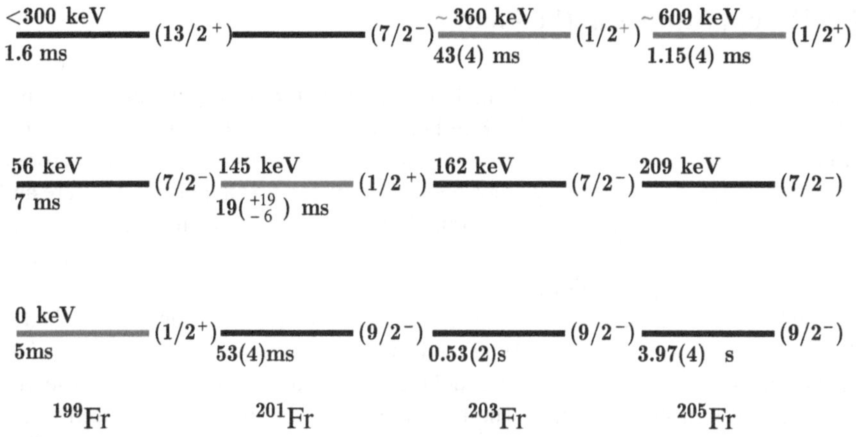

<300 keV (13/2⁺) 1.6 ms — (7/2⁻) ~360 keV 43(4) ms (1/2⁺) ~609 keV 1.15(4) ms (1/2⁺)

56 keV (7/2⁻) 7 ms — 145 keV 19($^{+19}_{-6}$) ms (1/2⁺) — 162 keV (7/2⁻) — 209 keV (7/2⁻)

0 keV (1/2⁺) 5ms — (9/2⁻) 53(4)ms — (9/2⁻) 0.53(2)s — (9/2⁻) 3.97(4) s

^{199}Fr ^{201}Fr ^{203}Fr ^{205}Fr

Fig. 1.1 Schematic illustration of the $(\pi 3s^{-1}_{1/2})_{1/2^+}$ proton intruder state in the neutron-deficient odd-A francium isotopes [7–9, 27]

well-known of the alkali metals. However, for reliable fundamental measurements to be performed, it is necessary to understand both the atomic and nuclear structure of this element, and for as many isotopes as possible. Measuring the most fundamental nuclear observables, such as nuclear spin, moments and change in mean-square charge radii, provides crucial information to test these fundamental theories.

The ability to study the neutron-deficient francium isotopes at the CRIS beam line offers the unique opportunity to answer the questions that have arisen from the study of the nuclear structure of isotopes in this region of the nuclear chart. As the isotopes above the Z = 82 shell closure become more neutron deficient, a decrease in the excitation energy of the $(\pi 1i_{13/2})_{13/2^+}$, $(\pi 2f_{7/2})_{7/2^+}$, $(\nu 1i_{13/2})_{13/2^+}$ and $(\pi 3s^{-1}_{1/2})_{1/2^+}$ states is observed. In ^{185}Bi (Z = 83) and ^{195}At (Z = 85), it is the $(\pi 3s^{-1}_{1/2})_{1/2^+}$ deformed intruder state that is the ground state [25, 26]. Recent radioactive decay measurements of ^{201}Fr and ^{203}Fr provide evidence for the presence of a $(\pi 3s^{-1}_{1/2})_{1/2^+}$ isomeric intruder state, suggesting this proton intruder state becomes the ground state in ^{199}Fr [7, 8]. The decrease in excitation energy of the $(\pi 3s^{-1}_{1/2})_{1/2^+}$ state as the $\nu 3p_{3/2}$ and $\nu 1i_{13/2}$ neutron orbitals are depleted can be seen in Fig. 1.1.

The intruder orbitals polarize the nucleus, creating significant deformation. From the study of the nuclear structure of the neutron-deficient francium isotopes towards ^{199}Fr (by measuring the spins, moments and change in mean-square charge radii of the ground and isomeric states), the ordering of the energy levels and shape of the nuclei can be investigated. In addition, decay spectroscopy on pure ground and isomeric states can be performed, to determine the lifetime and decay mechanisms of these isotopes. Using a combination of laser and decay spectroscopy, the key features in the structure of exotic nuclei can be revealed, furthering our understanding of the underlying nuclear force.

1.2 Recent Studies in the Field

The first optical measurements of the hyperfine structure of the francium isotopes $^{208-213}$Fr were performed at ISOLDE in 1978. This was initially with low-resolution laser spectroscopy [28], but later repeated with higher resolution [29]. The atomic transition 7s $^2S_{1/2} \rightarrow$ 7p $^2P_{3/2}$ was probed, allowing the first hyperfine structure and isotope shift measurements to be performed. Further measurements of the francium isotopes $^{207-213}$Fr and $^{220-228}$Fr were performed in 1985 [30–32] studying the same 7s $^2S_{1/2} \rightarrow$ 7p $^2P_{3/2}$ transition.

The first observation of the 7s $^2S_{1/2} \rightarrow$ 8p $^2P_{3/2}$ transition (the transition probed in the current work) was achieved in 1987 for the francium isotopes ^{212,213,220,221}Fr [33]. In both cases, the more exotic, neutron-deficient francium isotopes were not observed due to the low production rates. The laser atomic-beam spectroscopy (used in the measurement of the 7s $^2S_{1/2} \rightarrow$ 7p $^2P_{3/2}$ transition) along with the collinear fast-beam laser spectroscopy technique (employed to measure the 7s $^2S_{1/2} \rightarrow$ 8p $^2P_{3/2}$ transition) with fluorescence detection required high production rates (the most abundant isotope ^{212}Fr was produced at more than 10^9 atoms per second).

The CRIS technique, a combination of collinear laser spectroscopy and resonance ionization, was originally proposed by Kudriavtsev in 1982 [34] but the only experimental realization of the technique was not performed until 1991 on ytterbium atoms [35]. The standard collinear laser spectroscopy method overlaps an atomic (or ion) beam in a collinear geometry. This method reduces the Doppler broadening associated with the thermal motion of the atoms, resulting in a hyperfine structure that has a frequency range of MHz, instead of GHz. Thus, overlapping the accelerated atomic beam with laser light in a straight-line geometry increases the resolution of the technique. However, collinear laser spetcroscopy uses fluorescence detection, which requires higher production rates of the isotope under investigation. The process of resonant ionization (whereby the atomic beam is ionized) has a much higher detection efficiency than the fluorescence detection employed by standard collinear laser spectroscopy techniques. Resonantly ionizing the atomic bunch and detecting the individual ions allows isotopes with much lower production yields to be measured, in principle down to below 1 atom per second. Thus, the CRIS technique has the ability to perform hyperfine-structure measurements on the most neutron-deficient francium isotopes.

1.3 This Work

This thesis presents the hyperfine structure and decay studies of the neutron-deficient francium isotopes performed at the Collinear Resonance Ionization Spectroscopy (CRIS) experiment at ISOLDE. The nuclear observables of isotope shift, mean-square charge radii and magnetic dipole moment for the francium isotopes $^{202-207,211,220}$Fr measured with collinear resonance ionization spectroscopy are presented. Alongside these results are the radioactive-decay studies of the low-lying isomeric states of ^{202,204}Fr and ^{218m}Fr performed with the novel technique of laser assisted nuclear decay spectroscopy.

Chapter 2 outlines the theoretical considerations for collinear resonance ionization spectroscopy: the extraction of the nuclear observables from the hyperfine structure. Chapter 3 contains a brief introduction to nuclear structure: the theoretical models present in modern nuclear physics, the structure of the nucleus and the common radioactive-decay mechanisms. Chapters 4, 5 and 6 describe the experimental considerations for the production of ground state and isomeric ion beams. Chapter 4 describes the production of radioactive ion beams at the ISOLDE facility. Chapter 5 outlines the two techniques of collinear laser spectroscopy and resonance ionization spectroscopy combined at CRIS, in addition to the experimental components and requirements for collinear resonance ionization spectroscopy. Chapter 6 describes the experimental technique of laser assisted nuclear decay spectroscopy as well as providing a detailed description of the decay spectroscopy station. Chapter 7 presents the spectroscopic studies of the neutron-deficient francium isotopes performed at the CRIS beam line: collinear resonance ionization spectroscopy for hyperfine structure studies ($^{202-207,211,220,221}$Fr) and the complementary laser assisted nuclear decay spectroscopy for identification of ground and isomeric states (^{202,204}Fr), and decay spectroscopy of pure ion beams (^{202,204,218m}Fr). Chapter 8 discusses the results in the nuclear structure framework: the systematic evolution of the mean-square charge radii in the francium isotopes, the interpretation of the nuclear gyromagnetic (g-) factor and the systematics of the alpha-decay studies. Chapter 9 summarizes the results presented in this thesis and suggests complementary studies that would further benefit this work.

References

1. Hakala J et al (2008) Phys Rev Lett 101:052502
2. Cheal B et al (2010) Phys Rev C 82:051302
3. Verney D et al (2013) Phys Rev C 87:054307
4. Cheal B et al (2010) Phys Rev Lett 104:252502
5. Hornshoj P, Hansen P, Jonson B (1974) Nucl Phys A 230:380
6. Huyse M et al (1992) Phys Rev C 46:1209
7. Uusitalo J et al (2005) Phys Rev C 71:024306
8. Jakobsson U et al (2012) Phys Rev C 85:014309
9. Jakobsson U et al (2013) Phys Rev C 87:054320
10. Fedosseev VN et al (2012) Phys Scripta 85:058104
11. Weissman L et al (2002) Phys Rev C 65:024315
12. Stefanescu I et al (2007) Phys Rev Lett 98:122701
13. Van Roosbroeck J et al (2004) Phys Rev Lett 92:112501
14. Cheal B, Flanagan KT (2010) J Phys G 37:113101
15. Procter TJ et al (2012) J Phys: Conf Ser 381:012070
16. Gomez E et al (2006) Rep Prog Phys 69:79
17. Sheng D et al (2010) J Phys B At Mol Opt Phys 43:074004
18. Dzuba VA, Flambaum VV (2012) Phys Rev A 85:012515

19. Behr J et al (1993) Hyperfine Interact 81:197
20. Dzuba VA et al (1995) Phys Rev A 51:3454
21. Murphy M et al (2001) Mon Not R Astron Soc 327:1208
22. Murphy MT, Webb JK, Flambaum VV (2003) Mon Not R Astron Soc 345:609
23. Dzuba VA, Flambaum VV, Webb JK (1999) Phys Rev Lett 82:888
24. Dzuba VA et al (2002) Phys Rev A 66:022501
25. Davids CN et al (1996) Phys Rev Lett 76:592
26. Kettunen H et al (2003) Eur Phys J A 16:457
27. Uusitalo J et al (2013) Phys Rev C 87:064304
28. Liberman S et al (1978) C R Acad Sci Paris Ser B 286:353
29. Liberman S et al (1980) Phys Rev A 22:2732
30. Coc A et al (1985) Phys Lett B 163:66
31. Touchard F et al (1984) At Masses Fundam Constants 7:353–360
32. Bauche J et al (1986) J Phys B At Mol Opt Phys 19:L593
33. Duong HT et al (1987) EPL 3:175
34. Kudriavtsev Y, Letokhov V (1982) Appl Phys B 29:219
35. Schulz C et al (1991) J Phys B At Mol Opt Phys 24:4831

Chapter 2
Theoretical Considerations for Laser Spectroscopy

2.1 Hyperfine Structure

The coupling of the electronic angular momentum with the nuclear angular momentum leads to a substructure of the energy levels of the electronic orbitals, known as the hyperfine structure. For the case of ^{221}Fr, the hyperfine structure arising from the coupling of its nuclear spin I = 5/2$^-$ with the electronic orbitals 7p ^{2}S$_{1/2}$ and 8p ^{2}P$_{3/2}$ is shown in Fig. 2.1.

The electron has two components of angular momentum: the spin angular momentum, S, and the orbital angular momentum, L. In light atoms (usually Z < 30), these couple (known as LS coupling) to give the total electronic angular momentum J,

$$J = L + S, \tag{2.1}$$

where

$$L = \sum_i l_i \text{ and } S = \sum_i s_i. \tag{2.2}$$

In heavier atoms such a lead, bismuth and polonium, the individual orbital angular momentum, l_i, and spin angular momentum, s_i, combine to form a individual total angular momentum, j_i. These couple to form the total orbital angular momentum, J. This is known as jj coupling,

$$J = \sum_i j_i = \sum_i (l_i + s_i). \tag{2.3}$$

The coupling of the different projections of the spin and orbital angular momenta gives rise to the fine structure of the electronic orbitals. The total electronic angular momentum, J, in turn couples to the nuclear spin angular momentum, I, to give the total angular momentum, F,

$$F = I + J. \tag{2.4}$$

K.M. Lynch, *Laser Assisted Nuclear Decay Spectroscopy*,
Springer Theses, DOI: 10.1007/978-3-319-07112-1_2

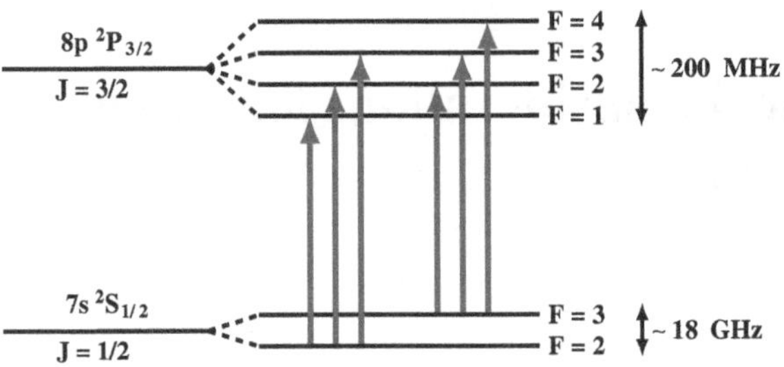

Fig. 2.1 Schematic illustration of the hyperfine structure of ^{221}Fr, where I = 5/2$^-$

When both I and J are greater then zero, degenerate hyperfine substates of the electronic orbitals are produced. The presence of the nuclear magnetic dipole moment or electric quadrupole moment raises the degeneracy of these substates, giving rise to a different energy for each level.

The perturbation of the hyperfine energy levels is given by [1, 2]

$$\frac{\Delta E}{h} = \frac{K}{2}A + \frac{3K(K+1) - 4I(I+1)J(J+1)}{8I(2I-1)J(2J-1)}B, \tag{2.5}$$

where $K = F(F+1) - I(I+1) - J(J+1)$. The hyperfine factors A and B are defined as

$$A = \frac{\mu_I B_e}{IJ}, \tag{2.6}$$

and

$$B = eQ_s\left\langle\frac{\partial^2 V_e}{\partial z^2}\right\rangle, \tag{2.7}$$

with μ_I the magnetic dipole moment of the nucleus, B_e the magnetic field of the electrons at the nucleus, Q_s the electric quadrupole moment, and $\langle\partial^2 V_e/\partial z^2\rangle$ the electric field gradient produced by the electrons.

The frequency, γ, at which the atomic transition between an upper and lower J level (J_u and J_l respectively) occurs is given by

$$\gamma = \nu + \alpha_{upper}A_{upper} + \beta_{upper}B_{upper} - \alpha_{lower}A_{lower} - \beta_{lower}B_{lower}. \tag{2.8}$$

Here, α and β are functions of the nuclear and atomic spin, as defined by

$$\alpha = \frac{K}{2}, \tag{2.9}$$

and

$$\beta = \frac{3K(K+1) - 4I(I+1)J(J+1)}{8I(2I-1)J(2J-1)}. \tag{2.10}$$

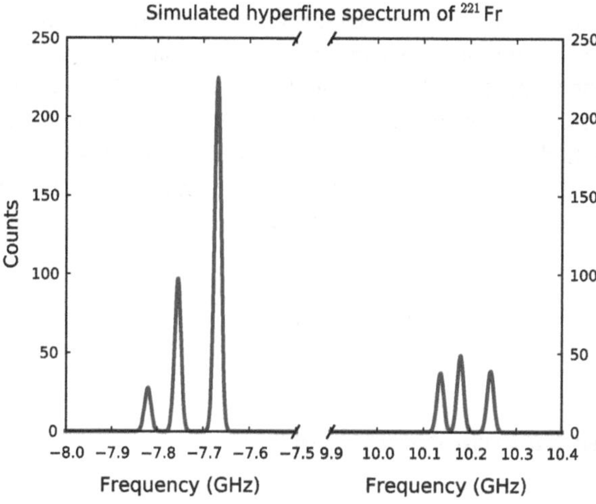

Fig. 2.2 Simulated hyperfine structure scan of ^{221}Fr. The lower 7p $^2S_{1/2}$ state splitting is ~18 GHz whereas the upper state splitting of 8p $^2P_{3/2}$ is significantly smaller at ~200 MHz

By fitting the hyperfine structure spectrum (such as one akin to Fig. 2.2) with a numerical routine such as χ^2-minimisation, the centroid frequency, ν, and the hyperfine factors $A_{upper,lower}$ and $B_{upper,lower}$ can be evaluated.

2.1.1 Nuclear Spin

For well resolved hyperfine structures, the nuclear spin of the isotope under investigation can often be determined from the relative frequencies of the atomic transitions, according to Eq. 2.5 if $J_{upper,lower} \leq 1/2$. In some cases, a spin can be immediately discounted due to the number of peaks present in the spectrum. For J = 1/2 to 1/2 or 0 to 1 transitions, the hyperfine structure does not provide enough information to determine the spin. For all other possible spins, χ^2 can be minimized and compared. The relative hyperfine-peak intensities determined by the weak-field angular coupling distribution can lead to a more significant difference in χ^2-minimisation [3]. However, careful experimental monitoring is required [4, 5].

2.1.2 Magnetic Dipole Moment

The magnetic dipole moment of the nucleus, μ_I, arises when the nuclear spin, I, is greater than zero. However, the magnitude of the magnetic moment when $I > 0$ can be vanishingly small, resulting in an A factor that leads to a hyperfine structure that is smaller than the natural line width of the state [6]. As shown in Eq. 2.6, the magnetic moment of the nucleus can be extracted from the hyperfine A factor. To the first order, the magnetic field at the nucleus due to the electrons is uniform along an isotopic chain, in the same way the electronic angular momentum, J, is. For a discussion on higher order corrections, see Sect. 2.1.4. This allows the magnetic moment of the isotope under investigation to be extracted from the known moment of another isotope of the element, using the ratio

$$\mu = \mu_{ref} \frac{IA}{I_{ref} A_{ref}}. \tag{2.11}$$

2.1.3 Electric Quadrupole Moment

The electric quadrupole moment, Q_s, can be determined in a similar fashion, for nuclei with $I > 1/2$ and $J > 1/2$. This is a result of the electric field gradient $\langle \partial^2 V_e / \partial z^2 \rangle$ produced by the electrons remaining constant along an isotopic chain. The quadrupole moment can be extracted from the magnetic moment and hyperfine B factor of a reference isotope, using the ratio

$$Q_s = Q_{s,ref} \frac{B}{B_{ref}}. \tag{2.12}$$

2.1.4 Hyperfine Anomaly

The Bohr-Weisskopf effect (BW) corrects the assumption that the nuclear magnetization is point-like [7]. For the $S_{1/2}$ and $P_{1/2}$ atomic states, the hyperfine interaction is affected by the non-uniformity of the magnetic field over the nuclear volume. For all other states, this effect is zero since there is virtually no overlap with the nucleus [8]. For heavy nuclei, the BW-effect is small, of the order of 1 % of the hyperfine A factor [9].

In addition, the Breit-Rosenthal effect (BR) corrects for the charge volume of the nucleus [10]. This effect is small for light nuclei but much larger for heavier nuclei, of the order of 20 % for Z=90 [9]. These two corrections reduce the hyperfine A factor to

$$A = A_{point-like}(1 - \epsilon_{BW})(1 + \epsilon_{BR}). \tag{2.13}$$

This leads to a modified expression for the magnetic moment,

$$\mu = \mu_{ref} \frac{IA}{I_{ref} A_{ref}} (1 + \Delta), \tag{2.14}$$

with,

$$\Delta = \frac{A\, g_{ref}}{A_{ref}\, g} - 1 \approx \epsilon - \epsilon_{ref}, \tag{2.15}$$

and the magnetic hyperfine anomaly defined as ϵ [11]. This can be calculated for nuclei whose nuclear gyromagnetic (g-) factors have been measured independent of laser spectroscopy, for example with nuclear magnetic resonance (NMR) spectroscopy [12].

The hyperfine anomaly can range from 10^{-5} to 1 % depending on the location of the isotope in the nuclear chart. For the francium isotopes, it is generally considered to be of the order of 1 % and is included as a contribution to the error [13].

2.2 Isotope Shift

The centre of gravity of the hyperfine structure (the centroid frequency) of one isotope relative to another, is shifted due to the difference in the structure of the two nuclei: their volume, shape, mass and charge radii. The isotopes shift, the shift of transition frequency of isotope A' compared to isotope A, can be written as

$$\delta\nu_{IS}^{A,A'} = \nu^{A'} - \nu^{A}. \tag{2.16}$$

The isotope shift can be evaluated as a linear combination of the mass shift and the field shift [14]

$$\delta\nu_{IS}^{A,A'} = \delta\nu_{MS}^{A,A'} + \delta\nu_{FS}^{A,A'}. \tag{2.17}$$

It arises (in part) due to the change in the mean-square charge radii between isotopes, associated with volume and shape changes.

2.2.1 Mass Shift

The mass shift component of the isotope shift originates from the recoil kinetic energy of a nucleus that has a finite mass. This shift can be calculated by

$$\delta\nu_{MS}^{A,A'} = M\frac{A' - A}{AA'}. \tag{2.18}$$

The M term, the mass factor, is dependent of the measured transition. For light nuclei, the mass shift is the significant contributing component of the isotope shift due to the $1/A^2$ dependence: the addition of a single neutron to a light nucleus has a much larger effect than adding one neutron to a heavy nucleus. The mass shift of an isomeric state relative to its ground state is zero as these nuclei have the same nuclear composition but are in a different nuclear state. However, with respect to the overall trend, the mass shift for isomers needs to be included. The remaining field shift is renamed the isomer shift, and gives the difference in centroid frequency between the ground state and the isomeric state.

The mass factor can be approximated in terms of a linear combination of the normal mass shift, K_{NMS}, and the specific mass shift, K_{SMS},

$$M = K_{NMS} + K_{SMS}. \tag{2.19}$$

The normal mass shift (NMS) is the contribution expected for a two-body system: the correction to the energy levels of the electrons relative to an infinitely heavy nucleus. This is always positive for the heavier isotope [15]. The normal mass shift is transition frequency, ν_{expt}, dependent and can be expressed as

$$K_{NMS} = \frac{\nu_{expt}}{1{,}822.888}. \tag{2.20}$$

The specific mass shift (SMS) is caused by the correlations between the electrons and can cause both a positive or negative shift. Calculation of this shift is non-trivial due to the evaluation of electron-correlation integrals. *Ab initio* calculations are only possible for nuclei with up to three electrons [16]. Heavier systems rely on large-scale many-body calculations, which are less accurate [17]. Alternatively, if experimental data is available, the specific mass shift can be determined by use of a King plot analysis [18]. This method is outlined in Sect. 2.2.4.

2.2.2 Field Shift

In heavy atoms, the isotope shift is dominated by the field shift: the shift in energy caused by the change in nuclear charge distribution as the nuclear content changes. This modifies the Coulomb interaction with the electrons. Over the nuclear volume, constant electron density is assumed and the perturbation in the electronic energy levels can be shown (to a first order approximation) to equal the mean-square charge radius, given by

$$\langle r^2 \rangle = \frac{\int_0^\infty \rho(r) r^2 \mathrm{d}V}{\int_0^\infty \rho(r) \mathrm{d}V}, \tag{2.21}$$

where $\rho(r)$ is the nuclear density. The field shift is sensitive to the change in the mean-square charge radius, as shown by relativistic calculations [19], thus the field shift is given by

$$\delta\nu_{FS}^{A,A'} = \frac{\pi a_0^3}{Z} \Delta|\psi(0)|^2 f(Z) \delta\langle r^2 \rangle^{A,A'}, \tag{2.22}$$

where $\Delta|\psi(0)|^2$ is the change in the probability density function of the electrons at the nucleus, a_0 is the Bohr radius, and $f(Z)$ a relativistic correction factor [20]. For isotopes of the same element, the atomic transition between s and p electrons yield the largest field shifts and are thus more sensitive to the difference in the mean-square charge radius, $\delta\langle r^2\rangle^{A,A'}$.

2.2.3 *Total Isotope Shift*

The isotope shift [21] can therefore be expressed as

$$\delta\nu^{A,A'} = M\frac{A'-A}{AA'} + F\delta\langle r^2\rangle^{A,A'}. \tag{2.23}$$

This separates the atomic and nuclear dependences: M and F are purely dependent on the atomic transitions and by comparison, $(A'-A)/AA'$ and $\delta\langle r^2\rangle^{A,A'}$ contain only the nuclear properties information.

2.2.4 *King Plot Analysis*

The atomic factors F and M can be evaluated by way of a King plot [18] if data is available for more than two stable isotopes whose charge radii were determined by other techniques (muonic x-rays or electron scattering). When no experimental data is available, the extraction of the atomic factors relies upon theoretical calculations, and introduces atomic-model dependence. The King plot analysis compares the isotope shifts of an element from two different atomic transitions. Equation 2.23 is first multiplied by the modification factor,

$$\mu_{A,A'} = \frac{AA'}{A'-A}, \tag{2.24}$$

for the transitions i and j, giving,

$$\begin{aligned} \mu_{A,A'}\delta\nu_i^{A,A'} &= M_i + \mu_{A,A'}F_i\delta\langle r^2\rangle^{A,A'}, \\ \mu_{A,A'}\delta\nu_j^{A,A'} &= M_j + \mu_{A,A'}F_j\delta\langle r^2\rangle^{A,A'}. \end{aligned} \tag{2.25}$$

After the elimination of $\mu^{A,A'}\delta\langle r^2\rangle$, the linear fit of the data $\mu^{A,A'}\delta\nu_j^{A,A'}$ against $\mu^{A,A'}\delta\nu_i^{A,A'}$ gives the straight line equation

$$\mu^{A,A'}\delta\nu_j^{A,A'} = \frac{F_j}{F_i}\mu^{A,A'}\delta\nu_i^{A,A'} + M_j - \frac{F_j}{F_i}M_i. \tag{2.26}$$

This has a gradient of F_j/F_i and an intercept of $M_j - (F_j/F_i)M_i$, whereby F and M can be evaluated for the transition of interest.

References

1. Schwartz C (1955) Phys Rev 97:380
2. Casimir HBG (1963) On the interaction between atomic nuclei and electrons. Freeman, San Francisco
3. Sato M (1955) Prog Theor Phys 13:405
4. Cheal B et al (2010) Phys Rev Lett 104:252502
5. Köster U et al (2011) Phys Rev C 84:034320
6. Mané E, Cheal B et al (2011) Phys Rev C 84:024303
7. Bohr A, Weisskopf VF (1950) Phys Rev 77:94
8. Grossman JS et al (1999) Phys Rev Lett 83:935
9. Büttgenbach S (1984) Hyperfine Interact 20:1
10. Rosenthal JE, Breit G (1932) Phys Rev 41:459
11. Barzakh AE et al (2012) Phys Rev C 86:014311
12. Panissod P (1986) Nuclear magnetic resonance, topics in current physics: microscopics models in physics. Springer, Berlin
13. Stroke HH et al (1961) Phys Rev 123:1326
14. King WH (1984) Isotope shifts in atomic spectra. Plenum, New York
15. Blaum K et al (2013) Phys Scripta 2013:014017
16. Yan ZC et al (2008) Phys Rev Lett 100:243002
17. Cheal B et al (2012) Phys Rev A 86:042501
18. King WH (1963) J Opt Soc Am 53:638
19. Bodmer AR (1953) Proc Phys Soc A 66:1041
20. Blundell SA et al (1985) Z Phys A 321:31
21. Cheal B, Flanagan KT (2010) J Phys G 37:113101

Chapter 3
Theoretical Considerations for Nuclear Decay Spectroscopy

3.1 The Nuclear Landscape

Approximately 3,000 known isotopes, and many more as-yet undiscovered, make up the nuclear landscape [1]. Schematically represented in Fig. 3.1, the 254 stable isotopes are shown in black, while those that undergo radioactive decay are colour coded. The four most common mechanisms of radioactive decay are β^- decay (blue), β^+ decay and electron capture (red), alpha decay (yellow) and spontaneous fission (green). The *magic numbers*, the proton and neutron numbers that display enhanced stability, are illustrated with black lines. The origin of these shell closures and the radioactive-decay mechanisms the nuclei undergo is discussed in the following chapter.

3.2 Theoretical Nuclear Models

The goal of theoretical nuclear physics is to ultimately describe the properties of nuclei in terms of mathematical models of their structure and constituent parts. At the macroscopic extreme is the *droplet model*, developed by Myers [2], describing the global properties of the nucleus. In contrast is the *shell model*, a quantum description of the individual constituents of the nucleus, developed by Goeppert-Mayer and Jensen [3, 4]. The following section briefly outlines the main properties of the nuclear models discussed in this thesis. A complete description of these models can be found in Refs. [5–7].

3.2.1 The Droplet Model

The basis of the droplet model, the *liquid drop model* of von Weizsäcker [8] takes a macroscopic approach to describe the properties of the nucleus. It treats the nuclear

K.M. Lynch, *Laser Assisted Nuclear Decay Spectroscopy*,
Springer Theses, DOI: 10.1007/978-3-319-07112-1_3

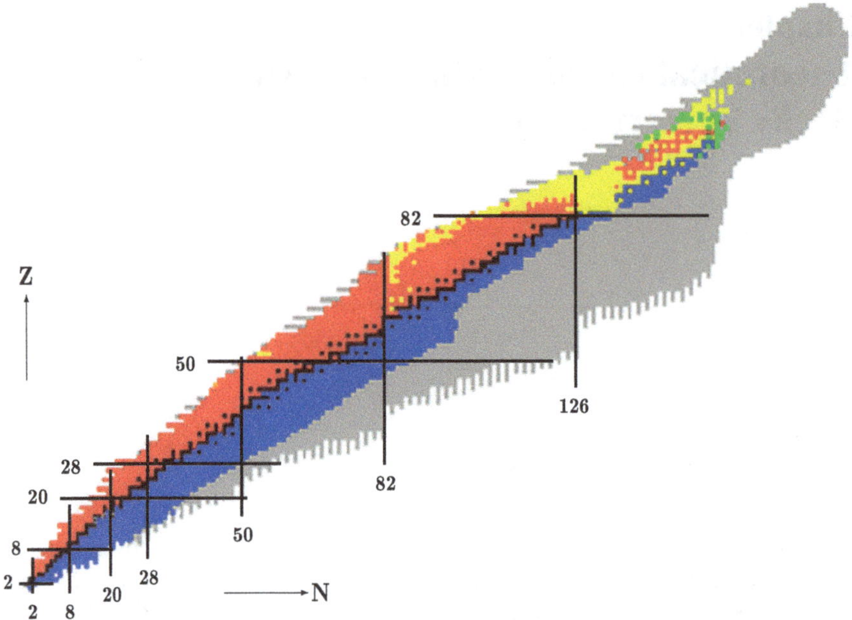

Fig. 3.1 The nuclear chart. The colour code represents the typical radioactive decay the isotope undergoes: β^- decay (*blue*), β^+ decay and electron capture (*red*), alpha decay (*yellow*) and fission (*green*). The *black squares* indicate stable isotopes and the *grey squares* denote predicted isotopes that have not yet been discovered, the 'terra incognita'. The *black lines* represent the shell closures, the 'magic numbers'

matter as an incompressible fluid to describe properties of the nucleus such as binding energy, mass and the fission barrier. It is the sum of five energies: the volume energy (the nuclear energy from the strong nuclear force), surface energy (to account for the surface nucleons interacting with fewer nucleons), Coulomb energy (the repulsive electrostatic interaction between protons), asymmetry energy (to account for the Pauli exclusion principle) and the paring energy (to correct for the pairing of valence protons and neutrons).

The modification to the standard liquid drop model to produce the *droplet model* is the inclusion of proton and neutron density distributions to minimise the total nuclear energy. This variation in nuclear density allows for surface diffuseness, in comparison to the sharp cut-off assumed in the liquid drop model. In addition to the macroscopic liquid drop model, the inclusion of an oscillating microscopic term accounts for fluctuations in the nuclear properties due to the nuclear single-particle level densities, the magic numbers. The inclusion of higher-order terms such as the surface diffuseness, surface curvature, nuclear compressibility and the relative neutron excess of heavier systems has increased the accuracy of this model when compared to experimental values [2]. Of the many bulk nuclear properties the droplet model can predict, the isotope shift, charge radii and binding energy are noted to be of specific interest for this work.

Dramatic improvements to this model came in the form of the *finite-range droplet model* (FRDM) which allowed nuclear-structure properties as well as nuclear ground-state masses to be calculated [9, 10].

3.2.2 The Shell Model

The shell model of nuclear theory describes the shell closures and increased stability exhibited by the proton and neutron magic numbers [5]. The underlying assumption of the shell model is that a single nucleon's motion is determined by the nuclear potential, a potential created by all the other nucleons in the nucleus. This Wood-Saxon potential, a modified version of the harmonic oscillator potential, is given by

$$V(r) = -\frac{V_o}{1 + \exp\left(\frac{r-R}{a}\right)}, \tag{3.1}$$

where $R = 1.25A^{1/3}$, $a = 0.524\,\text{fm}$ and $V_o \sim 50\,\text{MeV}$, empirically defined to be in agreement with the properties of known nuclei. Solving the Schrödingier equation in three dimensions removes the degeneracy of the major shells that arises from the simple harmonic oscillator potential. The resulting energy levels of the nuclear shells are shown in Fig. 3.2(Left). While the first few magic numbers of $Z, N = 2, 8, 20$ are reproduced, higher shell closures are not [6].

The inclusion of the spin-orbit interaction in the nuclear potential, $V_{so}(r)\ l \cdot s$, re-orders the energy levels and the magic numbers are entirely reproduced [4]. The resulting nuclear energy levels are shown in Fig. 3.2(Right). The spin-orbit interaction lifts the degeneracy of the $l > 0$ levels, whereby the split in energy for each l level increases with the magnitude of l.

The shell model dictates that only the single, unpaired, nucleon governs the properties of the nucleus. This results in a spin of j and a parity of $(-1)^l$ for a nucleus whose valence nucleon occupies the energy level $\text{n}l_j$.

3.2.3 The Deformed Models

Despite the shell model being the only broadly applicable microscopic model available, its use is limited. While shell-model calculations for light nuclei are possible (their proximity to closed shells limits the number of particles), the many valence nucleons present in heavier systems soon make calculations incredibly computer-intensive, if not impossible. To avoid the large valence space, and to address the absence of nuclear deformation in the shell model, two approaches have been taken.

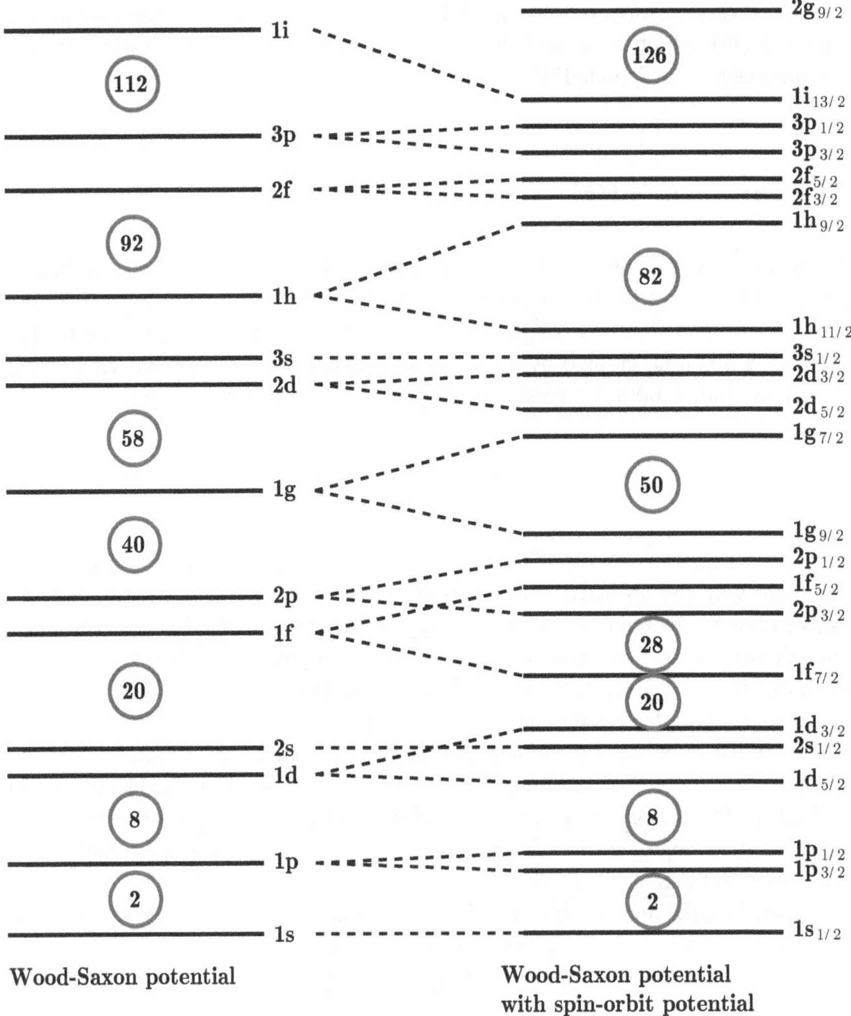

Fig. 3.2 The energy levels of the shell model. *Left* Calculated from the Wood-Saxon potential, as given in Eq. 3.1. *Right* Including the spin-orbit interaction. This lifts the degeneracy of the $l > 0$ levels and reproduces the magic numbers

3.2.3.1 The Deformed (Nilsson) Shell Model

The Nilsson model describes the single-particle states for deformed nuclei, the microscopic motion of single nucleons in a deformed potential. With the orbital angular momentum, l, and total angular momentum, j, no longer good quantum numbers, it is the projection of j on the symmetry axis, ω, and parity that is conserved.

The Nilsson model predicts the change in the shell energy levels and the shell gaps as nuclei undergo deformation.

3.2.3.2 The Collective Model

The collective model of Bohr and Mottelson is a macroscopic model that, similarly to the droplet model, describes the bulk motion and excitation of the nucleus. The addition of a deformation parameter of the nucleus allows rotational and vibrational excitations to be described.

3.3 Studying Nuclear Structure

3.3.1 Deformation of the Nucleus

The radius, R, of a deformed nucleus can be described by,

$$R(\theta, \phi) = R_{av}[1 + \beta Y_{20}(\theta, \phi)], \tag{3.2}$$

where $R_{av} = R_0 A^{1/3}$ and $Y_{20}(\theta, \phi)$ is the second-order spherical harmonic. This equation is independent of ϕ which results in the symmetry of the nucleus about one axis. The deformation parameter, β, is defined as,

$$\beta = \frac{4}{3}\sqrt{\frac{\pi}{5}}\frac{\Delta R}{R_{av}}, \tag{3.3}$$

where ΔR is the difference between the two axes of the deformed nucleus. When $\beta > 0$, the nucleus is elongated along the axis of symmetry (relative to θ) and has the form of a *prolate* spheroid. In contrast, when $\beta < 0$, the nucleus is compressed along its symmetry axis and becomes an *oblate* spheroid. Figure 3.3 shows the two quadrupole shapes of the nucleus: (a) prolate and (b) oblate [6]. Static prolate shapes, typically present when $Z \sim 80$ and $150 < A < 190$, are more common than oblate shapes across the nuclear chart. Oblate shapes only occur when Z or N is near the end of a major shell (for example, in the mercury isotopes that are discussed in Sect. 8.1) [7].

The deformation of the nucleus, β, (observed in the body-fixed frame) can be measured by the intrinsic quadrupole moment,

$$Q_0 = \frac{3}{\sqrt{5\pi}} R_{av}^2 Z\beta(1 + 0.16\beta). \tag{3.4}$$

For well-deformed ground states, the intrinsic quadruple moment is related to the spectroscopic quadrupole moment, Q_s, by,

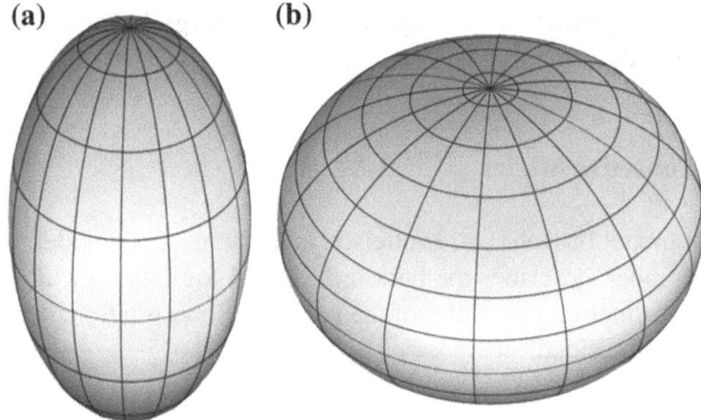

Fig. 3.3 **a** Prolate and **b** oblate nuclear deformations. Images courtesy of Wikimedia Commons [11, 12]

$$Q_s = \frac{3\langle K^2\rangle - I(I+1)}{(I+1)(2I+3)} Q_0, \tag{3.5}$$

where K is the projection of the total spin on the nuclear axis [13]. In the strong coupling limit, it is convention to set $K = I$. Hyperfine structure studies measure the spectroscopic quadruple moment of the nucleus, providing the time-averaged static deformation of the nucleus.

The mean-square charge radius describes both the static and dynamic deformation of the nucleus. It can be approximated as a multipole (λ) expansion of

$$\langle r^2\rangle \approx \langle r^2\rangle_0\Big(1 + \frac{5}{4\pi}\sum_{\lambda}\langle\beta_\lambda^2\rangle\Big), \tag{3.6}$$

where $\langle r^2\rangle_0 = \frac{3}{5}r_0^2 A^{2/3}$ [13]. Due to the square of the deformation parameter β_λ, information on the prolate or oblate nature of the deformation is lost. Thus the charge radius provides only the magnitude of the deformation, both static and dynamic. The dynamic deformation arises from rotational excitations of a deformed nucleus, vibrational excitation of a spherical nucleus or changes in the surface diffuseness. Additional information on the nature of the deformation can be gained from the comparison with the charge radii from theoretical models, such as the droplet model described in Sect. 3.2.1.

3.3.2 Shape Coexistence in the Region of ^{186}Pb

Shape coexistence is the phenomenon of distinctly different nuclear shapes occurring at similar binding energies (within a couple of MeV). The energy eigenstates of two (or more) nucleon arrangements with different electric quadrupole properties can invert with the addition (or removal) of a nucleon. This results in a dramatic change in the shape of the ground state nucleus. The presence of shape coexistence is a result of the interplay between the stabilising effect of the closed shells (to maintain a spherical shape) and the residual interaction between protons and neutrons. Thus, this phenomenon is often observed near singly closed (proton or neutron) shell regions near the middle of the shell of the other nucleon [14].

Several spectroscopic characteristics exist that can indicate the presence of shape coexistence. Nuclear deformation can be directly measured by the diagonal E2 matrix elements (from multi-step Coulomb excitation) and B(E2) values (from lifetime measurements). In addition, changes in nuclear masses (from two-proton separation energies) and mean-square charge radii (from isotope and isomer shifts) can indicate the presence of shape coexistence. Alpha decay patterns (from hindrance factors) can also be used to identify the change in nuclear configuration [14].

The francium isotopes are located five protons above the lead isotopes at the shell closure of Z = 82 [15]. The occupation of the shell-model levels in francium are shown in Fig. 3.4. The filled circles denote ^{202}Fr, while the open circles illustrate the filled neutron states of ^{213}Fr. The placement of ^{204}Fr with respect to the neighbouring magic numbers of Z = 82 and N = 126 can be seen in Fig. 3.5. The colour code represents the binding energy per nucleon (BE/A) of each isotope.

One region of the nuclear chart where shape coexistence occurs is identified by the black ellipse in Fig. 3.5, centred around ^{186}Pb. This nucleus is located at the shell closure of Z = 82 and mid-shell at N = 104, halfway between N = 82 and 126. In ^{186}Pb, a triplet of 0^+ states have been discovered, all within an excitation energy of less than 700 keV [16]. The three states, with assumed spherical, prolate and oblate nuclear shapes, can be seen in Fig. 3.6. This figure is reproduced from Ref. [16] and illustrates the theoretical deformation energy surface. Three minima can be observed, corresponding to the three low-lying states of ^{186}Pb, each with a different nuclear shape.

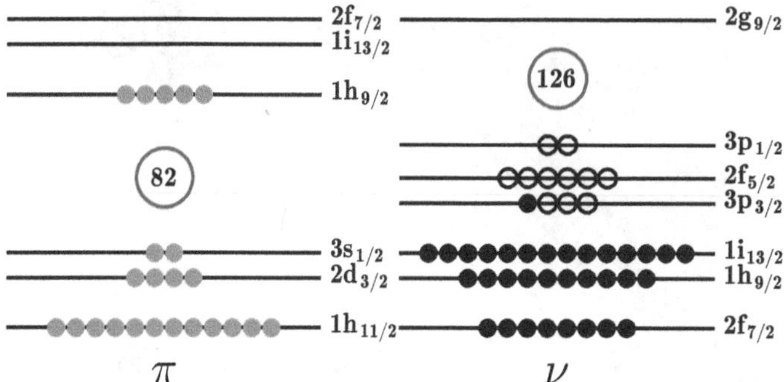

Fig. 3.4 The occupation of the shell model levels for ^{202}Fr (N = 115, *filled circles*) to ^{213}Fr (N = 126, *open circles*)

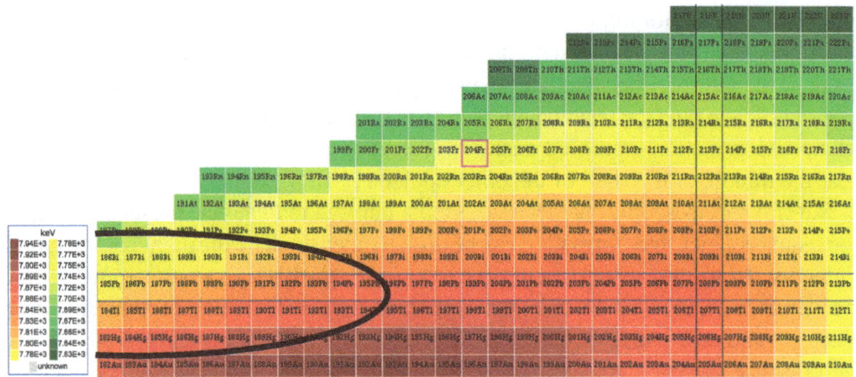

Fig. 3.5 The location of francium (Z = 87) in the nuclear chart. The colour code represents the binding energy per nucleon (BE/A) of each isotope [15]. The *circled* isotopes denote the region of shape coexistence, centred around ^{186}Pb (Z = 82, Z = 104)

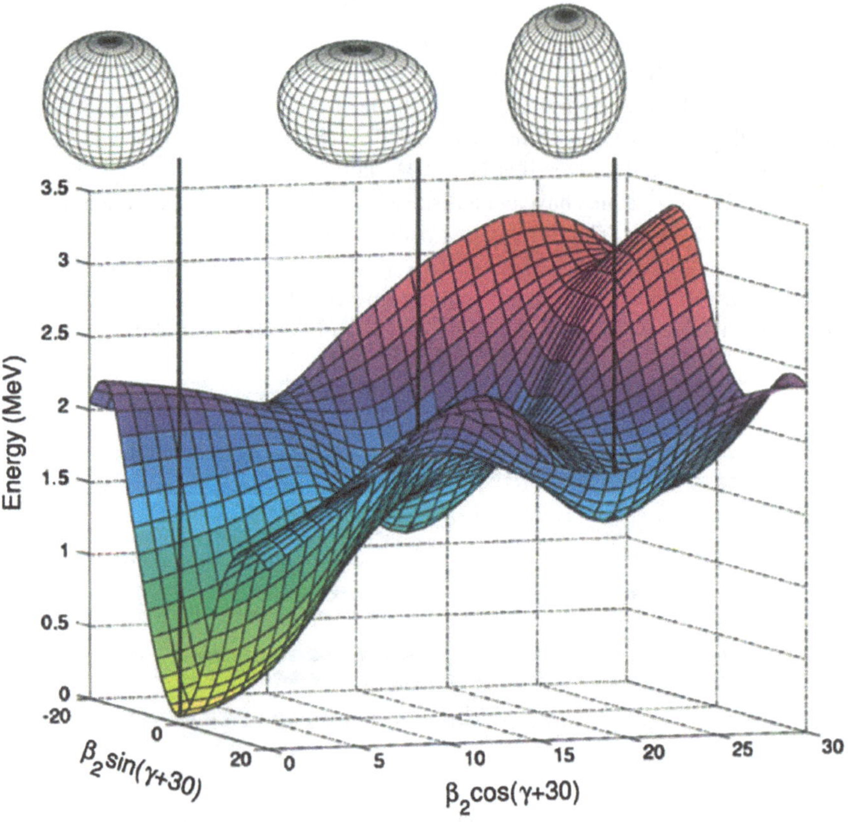

Fig. 3.6 The three low-lying states of ^{186}Pb. The theoretical calculation of the deformation energy surface of ^{186}Pb predicts three minima corresponding to spherical, oblate (2p-2h excitation) and prolate (4p-4h excitation) shapes. Reproduced from Ref. [16]

At mid-shell, there is a maximum number of valence protons and neutrons that can be involved in particle excitations that leads to deformation. The first 0^+ state results from the coupling of the paired neutrons and protons in the even-even ^{186}Pb nucleus. The second 0^+ state, at an excitation energy of 532 keV is associated with a proton-pair excitation across the Z = 82 shell gap, resulting in a 2-particle 2-hole configuration coupling to 0^+. The third 0^+ state is believed to arise from 4-particle 4-hole coupling, at an excitation energy of 650 keV.

3.3.3 Proton Intruder States

Intruder levels are caused by the attractive interaction between proton and neutron states. With a sufficient number of valence protons and neutrons, which can result from particle-hole excitations across a shell closure, the attractive proton-neutron interaction creates deformed intruder states. This interaction is greater than the interaction between non-intruder levels, resulting in a decrease in the excitation energy of the intruder state. This lowering of the energy increases linearly with the number of neutrons, reducing the shell effect as the mid-shell is approached [7].

In heavy nuclei, such as lead and francium, there is a large neutron excess. This results in the excited valence protons in the intruder state occupying nuclear shells with a large quantum overlap with those of the valence neutrons. This enhances the proton-neutron interaction, and leads to intruder states and shape coexistence in this region of the nuclear chart.

3.4 Radioactive Decay Mechanisms

3.4.1 Alpha Decay

Many heavy nuclei (with $Z > 82$, or $Z < 82$ but $A > 100$) radioactively decay through emission of an alpha particle. This tightly bound particle is the nucleus of $^4_2$He, and is emitted with a high kinetic energy due to its high binding energy.

$$^{A}_{Z}X_N \rightarrow \, ^{A-4}_{Z-2}X'_{N-2} + \alpha \tag{3.7}$$

Alpha decay is the result of Coulomb repulsion. As the nucleus becomes larger, the repulsive Coulomb force increases with size, of the order of Z^2. This is larger in magnitude than the stabilising binding energy of the nucleus, which increases with A, and the disruptive Coulomb force causes an emission of a high-energy alpha particle.

The angular momentum of the emitted alpha particle can range from $|I_i - I_f|$ to $|I_i + I_f|$ when a nucleus decays from an initial state with angular momentum I_i to

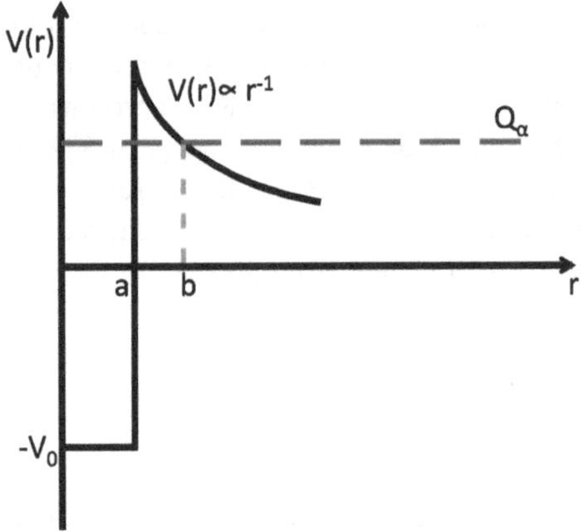

Fig. 3.7 The potential energy, $V(r)$, of an alpha particle as a function of separation from the nucleus, r. An alpha particle with energy $(V_0 + Q_\alpha)$ will quantum mechanically tunnel through the potential barrier from a to b. See text for details

a final state I_f. This angular momentum, l_α, is orbital in nature due to the coupling of the proton-neutron pairs in the alpha particle to spin zero. The change in parity of alpha emission is $(-1)^{l_\alpha}$, leading to the selection rule for alpha decay. For an allowed transition: if there is a change in parity from one state to another, l_α must be odd; if the parities are the same, l_α must be even.

The decay of a nucleus through alpha emission can result in the daughter nucleus populating an excited state. This is called the *fine structure* of alpha decay, a different description to the fine structure of atomic levels, as discussed in Chap. 2. The relative intensity of population of the excited states in the daughter nucleus is given by alpha-branching ratios. These present the relative intensities of alpha decay to each daughter state, highly dependent on the wave function of the initial and final states, and angular momentum considerations.

The Q-value is the total energy released by alpha decay and is given by,

$$Q_\alpha = (m_X - m_{X'} - m_\alpha)c^2 \tag{3.8}$$

where m is the mass of the parent (X), daughter (X') and alpha particle (α), respectively. Alpha emission occurs spontaneously for $Q_\alpha > 0$. The kinetic energy of the alpha particle is then given by

$$E_\alpha = \frac{Q_\alpha}{1 + m_\alpha/m_X}. \tag{3.9}$$

Figure 3.7 illustrates a simplistic view of the potential energy, $V(r)$, of an alpha particle, relative to its separation from the nucleus, r, with radius a. When $r < a$, the potential energy is represented as a square well of depth V_0. For $r > a$ the potential energy is determined by Coloumb repulsion and is proportional to r^{-1}. Alpha particle emission occurs when the (preformed) alpha particle quantum mechanically tunnels through the potential barrier from a to b.

The Geiger-Nuttall law of alpha decay states that radioactive nuclei that emit an alpha particle of high energy will have a short lifetime, with the converse holding true. A plot of $\log(t_{1/2})$ against Q_α will give a smooth curve for even-even nuclei [6].

3.4.2 Beta Decay

Beta decay is a three-body process whereby the nucleus decays through conversion of nucleon and emission of two leptons. In β^- decay, the nucleus converts a neutron to a proton and an electron and anti-neutrino is emitted.

$$\beta^-\text{decay: } {}^A_Z X_N \rightarrow {}^{A}_{Z+1}X'_{N-1} + e^- + \overline{\nu} \tag{3.10}$$

Similarly in β^+ decay, a constituent proton is converted to a neutron and the nucleus emits a positron and a neutrino.

$$\beta^+\text{decay: } {}^A_Z X_N \rightarrow {}^{A}_{Z-1}X'_{N+1} + e^+ + \nu \tag{3.11}$$

Beta decay allows an unstable nucleus to traverse along the isobaric line of constant A towards stability [6]. There are several exotic decay mechanisms for beta decay such as double-beta decay, beta-delayed (one or two) nucleon emission and beta-delayed fission, which are discussed in detail elsewhere [6, 17, 18].

3.4.3 Gamma Decay

Gamma decay has been observed in all nuclei that have bound excited states, those with $A > 5$. It occurs when an excited state decays to a lower energy excited state or ground state. It does this by emitting a photon of gamma radiation, with an energy equal to the difference in energy between the two states, often between 0.1 and 10 MeV. Alpha or beta decay is often followed by gamma decay, as charged-particle emission typically leaves the nucleus in an excited state. Such a decay scheme is shown in Fig. 3.8 for ^{60}Co. The emission of beta particles into excited states of ^{60}Ni is followed by gamma decays of energies 1,173.2 and 1,332.5 keV [15].

The emitted electric (E) or magnetic (M) radiation has an associated multipole order of 2^L where the angular momentum $L = 1$ for dipole, 2 for quadrupole, 3 for octupole radiation, and so forth. The parity π of the emitted electric or magnetic radiation is

$$\pi(EL) = (-1)^L, \text{ or, } \pi(ML) = (-1)^{L+1}, \tag{3.12}$$

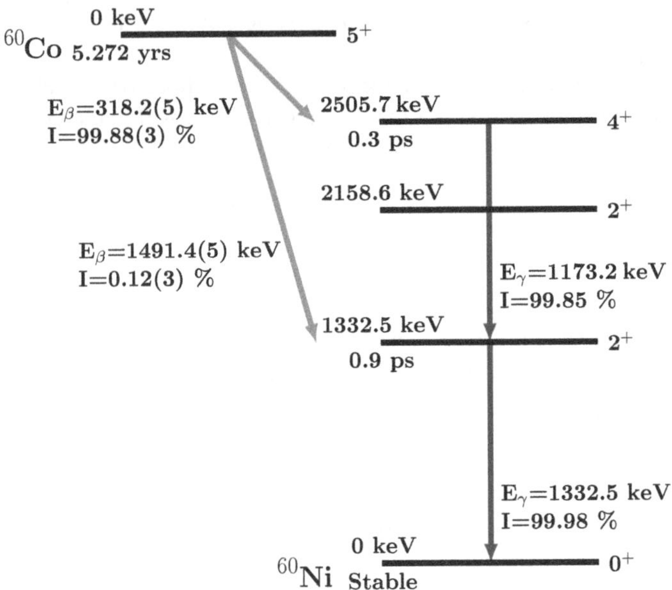

Fig. 3.8 The radioactive decay of ^{60}Co. Beta particle emission to excited states in ^{60}Ni is followed by the characteristic gamma decays of energies 1,173.2 and 1,332.5 keV [15]

respectively. Similar to the alpha-decay transition rules, the angular momentum of radiation emitted from a transition (from initial state L_i to final state L_f) can range from $|L_i - L_f|$ to $|L_i + L_f|$, as long as L $\neq$ 0. The electric or magnetic character of the emitted radiation is determined by the change in parity of the transition. For no change in parity between the initial and final states, if the angular momentum L is even, the radiation will be electric, if odd the radiation will be magnetic. Similarly, if a change in parity is observed, the radiation is electric if L is odd, magnetic if L is even.

3.4.4 Internal Conversion

A competing process to gamma decay is internal conversion, where the nucleus de-excites by transferring its energy to an atomic electron. The electromagnetic multipole field interacts with the atomic electrons, transferring its energy and an electron is emitted with a discrete amount of energy. The kinetic energy of the electron is given by,

$$T_e = \Delta E - BE, \tag{3.13}$$

the difference in the transition energy, ΔE, less the binding energy, BE, of the electron. The electron vacancy (usually located in one of the inner shells) is filled by an electron in a higher-lying shell, emitting a characteristic X-ray as it transitions.

3.5 Isomerism

An isomeric state (denoted by the exponent m) is an long-lived excited state of the nucleus. Typical half-lives for gamma emission are less than 10^{-12} s, but isomeric states can be significantly longer, from less than 10^{-7} s (in the case of ^{139m}Eu [19]) to 31(1) years (^{178m2}Hf [20]) or longer. The isomer of ^{180}Ta has not yet been observed to decay, resulting in an adopted half-life estimate of over 1.2×10^{15} years [21].

For the purpose of this work, isomeric states are considered to be states that can be produced and extracted on-line in the form of a low-energy beam for spectroscopy studies (typically $t_{1/2} \geq 1$ ms).

References

1. Erler J et al (2012) Nature 486:509
2. Myers WD (1977) Droplet model of atomic nuclei. Plenum, New York
3. Mayer MG (1948) Phys Rev 74:235
4. Mayer MG, Jensen JHD (1955) Elementary theory of nuclear shell structure. Wiley, New York
5. Heyde KLG (1994) The nuclear shell model. Springer, Berlin
6. Krane KS (1988) Introductory nuclear physics. Wiley, New York
7. Casten RF (2000) Nuclear structure from a simple perspective. Oxford University Press, Oxford
8. von Weizsäcker CF (1935) J Phys 96:431
9. Möller P et al (1988) Atomic Data Nucl Data Tables 39:225
10. Möller P et al (1995) Atomic Data Nucl Data Tables 59:185
11. http://en.wikipedia.org/wiki/file:prolatespheroid.png
12. http://en.wikipedia.org/wiki/file:oblatespheroid.png
13. Poenaru D, Greiner W (1996) Handbook of nuclear properties. Oxford University Press, Oxford
14. Heyde K, Wood JL (2011) Rev Mod Phys 83:1467
15. http://www.nndc.bnl.gov/nudat2/
16. Andreyev AN (2000) Nature 405:430
17. Elliott SR, Vogel P (2002) Ann Rev Nucl Part Sci 52:115
18. Hall HL, Hoffman DC (1992) Ann Rev Nucl Part Sci 42:147
19. Cullen DM et al (2011) Phys Rev C 83:014316
20. Achterberg E et al (2009) Nucl Data Sheets 110:1473
21. Wu SC, Niu H (2003) Nucl Data Sheets 100:483

Chapter 4
Experimental Setup at ISOLDE

Radioactive ion beams of over 70 elements, and 1,000 different isotopes, are produced at the Isotope Separator On-Line DEvice (ISOLDE) facility at CERN (Geneva, Switzerland) from the reactions of high-energy protons on thick targets [2]. The ISOLDE facility performed its first experimental test on 16th October 1967, with the results published in Physics Letters in 1969 [3]. Initially, 600 MeV protons were delivered to ISOLDE from the CERN proton synchrocyclotron (SC); and a decade later, after the completion of the SC Improvement Program (SCIP), the proton beam intensity on the ISOLDE target was increased by a factor of 10^2, allowing shorter-lived and therefore more exotic nuclear species to be investigated. It was during this time that the first optical measurements of francium were performed. In 1978, Liberman identified the 7s $^2S_{1/2} \rightarrow$ 7p $^2P_{3/2}$ atomic transition, performing hyperfine structure and isotope shift measurements first with low-resolution [4] and later with high-resolution laser spectroscopy [5]. The wavelength of this transition $\lambda(D_2) = 717.97(1)$ nm was in excellent agreement with the prediction of Yagoda [6], made in 1932 before francium was even discovered.

Further measurements of francium, the heaviest alkali metal, followed in the next decade. High-resolution optical measurements were performed on both the 7s $^2S_{1/2}$ $\rightarrow$ 7p $^2P_{3/2}$ atomic transition [7–9], as well as the 7s $^2S_{1/2} \rightarrow$ 8p $^2P_{3/2}$ transition [10], along with transitions into high-lying Rydberg states [11]. The volume of output in these years can be attributed to not only the expertise of the scientists involved, but the high yield of francium production at ISOLDE. The on-line production of ^{212}Fr (with a 20 min half life) is of the order of 10^9 atoms per second, comparable to the amount obtained from processing 10 tons of uranium ore in one second [2].

Many upgrades to the ISOLDE facility took place in the following years. Most notable was the movement of the whole facility to receive protons from the CERN proton synchrotron booster (PS Booster) instead of the old SC. The construction of the new site started in 1990, and by 26th June 1992 the first experiments at the relocated ISOLDE took place [12]. The PS Booster now delivers a 1.4 GeV pulsed proton beam, with each pulse separated by 1.2 s. This pulsed beam can contain up to 45 pulses each super-cycle, with ISOLDE taking pulses at irregularly spaced intervals throughout the cycle, at the average intensity of up to 2 μA (10^{13} protons per second).

K.M. Lynch, *Laser Assisted Nuclear Decay Spectroscopy*,
Springer Theses, DOI: 10.1007/978-3-319-07112-1_4

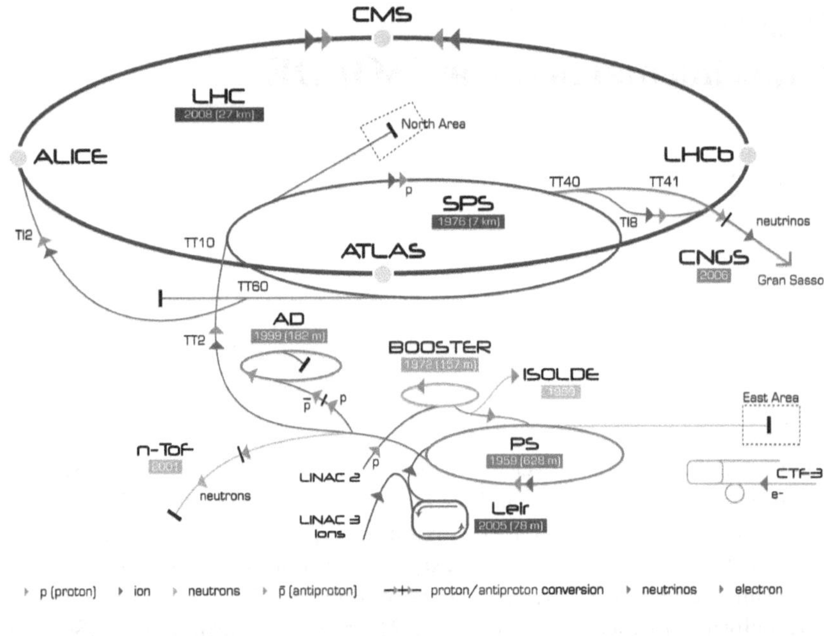

Fig. 4.1 Layout of the CERN accelerator complex. Protons from the PS Booster are delivered to the ISOLDE facility (shown in green). Image courtesy of Ref. [1]

The location of the ISOLDE facility in the CERN accelerator complex is shown in Fig. 4.1. ISOLDE shares the proton-beam delivery with the other CERN experiments, for example Neutrons Time of Flight (n-ToF) [13], the Antiproton Decelerator (AD) [14], CERN Neutrinos to Gran Sasso (CNGS) [15] and perhaps most prominently the Large Hadron Collider (LHC) [16]. In addition, there are several fixed-target experiments located in the PS test beam facility in the East Area and the SPS test beam facility in the North Area to which the protons are delivered. The proton bunches are delivered according to a super cycle of 30–45 pulses, with ISOLDE typically receiving ~50 % of the protons during usual running conditions.

4.1 Radioactive Beam Production

There are two families of radioactive ion beam facilities, producing beams using two general methods: isotope separation on line [17] and in-flight separation [18].

4.1.1 Isotope Separation on Line

ISOLDE uses the isotope separator on line (ISOL) technique [17]. In ISOL, exotic isotopes are typically created by light ion induced reactions on a thick target. This target is thick enough to stop the recoiling reaction products, which are extracted from the target material (by effusion and diffusion) and ionized in an ion source. The ions are mass separated using dipole magnets and accelerated to the required energy. This allows for a high quality beam of the chosen isotope, but is limited in delivering short-lived species by the time taken for the ion to be extracted from the target matrix [20].

This method is used by many facilities around the world (see Fig. 4.2), for example Louvain-la-Neuve in Belgium (LISOL) [21], TRIUMF in Canada (ISAC) [22] and GANIL in France (SPIRAL1) [23]. Jyväskylä in Finland (IGISOL) [24] uses a thin (foil) target in place of a thick target and stops the reaction products in a buffer gas. The future facility of SPIRAL2 at GANIL will use the ISOL technique, as well as the next generation European ISOL facility EURISOL [25].

4.1.2 In-Flight Separation

The in-flight separation (IFS) method of producing radioactive ion beams is used at facilities such as GANIL in France (LISE-3), NSCL in the USA (A1200) [26], RIKEN in Japan (RIPS) [27] and GSI in Germany (FRS) [28] (see Fig. 4.2). This technique uses a heavy-ion beam impacting onto a thin target, exploiting the kinematics of the reactions, namely that the reaction products are scattered in the forward direction. The fragments have similar energies to that of the incoming beam due to the thin target, negating the requirement of re-acceleration. As the loss in energy of the required reaction products are kept below $\Delta E/E < 10\,\%$, the momentum distribution of these fragments is narrow, and they can therefore be selected using the optics of a momentum-selective spectrometer [29].

Despite this process allowing for isotopes with shorter half-lives to be produced, and at a higher energy (allowing for the option of secondary nuclear reactions to take place), the optical quality of the beam is the major limitation. The high energy and energy spread of the ion beam are currently not suitable for precision experiments such as laser spectroscopy. However, recent advances in gas-catcher technology at NSCL (FRIB) [30] and RIKEN (SLOWRI) [31] will provide thermalized projectile fragments that will allow for such measurements to be performed. The IFS technique will be used at the upgraded facilities of A1900 at NSCL [32], Big-RIPS in Japan [33] as well as at Super-FRS (and the future FAIR facility) in Germany [34].

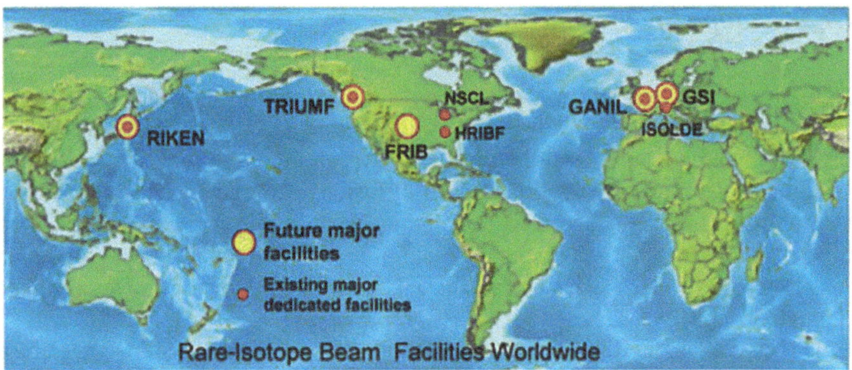

Fig. 4.2 Major radioactive ion beam facilities worldwide. RIKEN: Rikagaku Kenkyusho (Institute of Physical and Chemical Research, Japan); TRIUMF: Tri-University Meson Facility; FRIB: Facility for Rare-Isotope Beams; NSCL: National Superconducting Cyclotron Laboratory; HRIBF: Holifield Radioactive Ion Beam Facility; GANIL: Grand Accélérateur National d'Ions Lourds; ISOLDE: Isotope Separator OnLine DEvice; GSI: Gesellschaft für Schwerionenforschung. Image courtesy of Ref. [19]

4.2 Isotope Production, Ionization and Separation

4.2.1 Isotope Production

The protons delivered to ISOLDE from the PS Booster impinge onto one of over 25 different target materials [35]. The radioactive recoils produced from these reactions are extracted and ionized in one of three ways: by surface ionization, by plasma ionization, or by laser excitation. Ions are then accelerated by a high voltage, mass separated and delivered to the ISOLDE experimental hall where a range of experiments are performed. The layout of the ISOLDE facility can been seen in Fig. 4.3.

"An ISOLDE target is, in reality, a small chemical factory..." [2] where bombardment of protons onto the target causes spallation, fission and fragmentation to occur. Combining these reactions with the large array of target-ion-source combinations available, nuclei at the very edges of stability can be produced and radioactive beam experiments performed.

4.2.2 Isotope Ionization

Figure 4.4 shows the schematics of the three methods in which radioactive isotopes are ionized in the target at ISOLDE. The first, and simplest, method is surface ionization, shown in Fig. 4.4a. This setup consists of a transfer line (e.g. a tantalum metal tube) which the atoms diffuse through. Collision of the atom with a heated surface

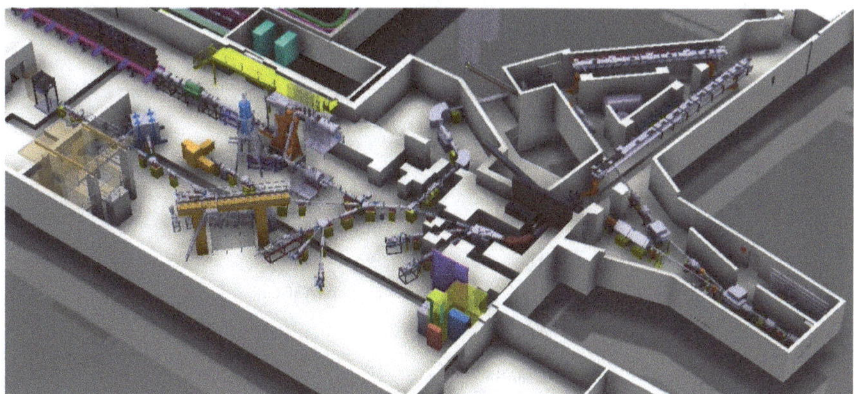

Fig. 4.3 Layout of the ISOLDE facility. Image courtesy of Ref. [36]

causes a transfer of electrons: the atom becomes a negatively or positively charged ion. For a negative ion, the atom's electron affinity is required to be greater than 1.5 eV due to the low work function of the surfaces used. For positive ion creation, the atom's ionization potential needs to be less than 7 eV. For positively charged ions, the efficiency of ionization by surface ionization, $\epsilon_{surface}$, is given by

$$\epsilon_{surface} = \frac{n_i}{n_i + n_0}, \tag{4.1}$$

where $n_{i,0}$ is the ion or neutral density, respectively. This formula has a strong dependence on the surface temperature and difference between the work function of the surface and the ionization potential of the atom. For this reason, metals such tantalum, tungsten or rhenium are used, due to their high work functions of 4.19 eV, 4.53 eV and 5.1 eV respectively. The transfer line is heated to 2,400° C, in order to optimize ionization [37].

The second production method uses a plasma ion source, when (due to a high ionization potential) the isotope of interest cannot be surface ionized with a sufficient efficiency. An ionized plasma of noble gases (such as argon and xenon) is achieved by accelerating electrons from the transfer line to extraction electrode by supplying a voltage of 130 V to the anode. An additional magnet (labeled SRCMAG on Fig. 4.4b) provides a magnetic field to confine the plasma. If the isotope of interest is a noble gas, the transfer line is cooled (instead of heated) to limit the less volatile elements from being transported along the transfer line, and reduce isobaric contaminants [36].

The third method of ion production applies atomic physics to isolate the elements of interest. The Resonance Ionization Laser Ion Source (RILIS) at ISOLDE separates out the atoms of interest using their unique atomic properties [38], as shown in Fig. 4.4c. Again the surface ion source is used, however ions are created not only by surface ionization but also by step-wise resonance photo-ionization with lasers. The laser beams are overlapped with the transfer line, resonantly exciting the atom via atomic intermediate states, until an electron is brought beyond the ionization

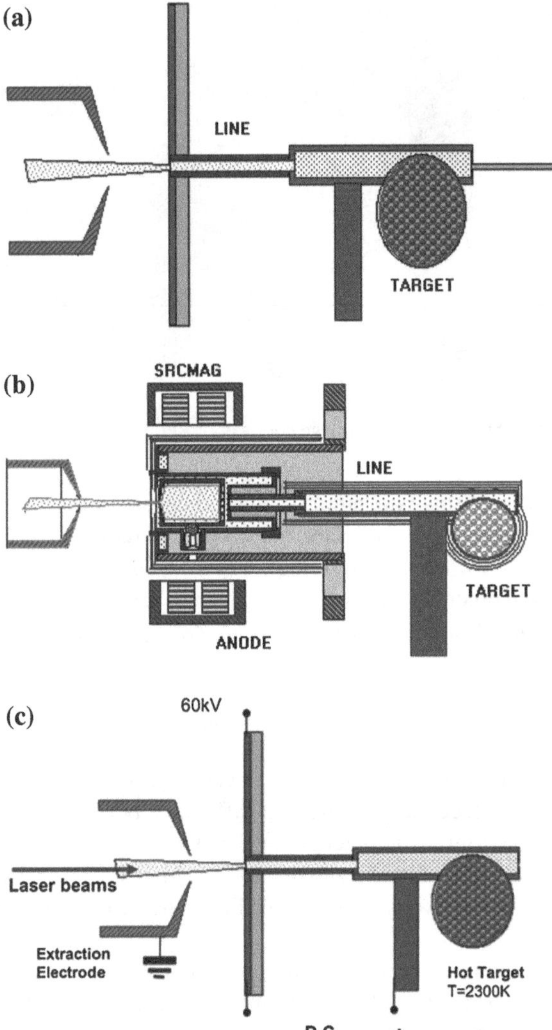

Fig. 4.4 The three types of ion source at ISOLDE. **a** Surface ion source **b** Plasma ion source **c** Laser ion source Image courtesy of Ref. [36]

threshold. The ionization of an excited atom can be achieved non-resonantly by exciting directly into the continuum or resonantly through an auto-ionizing or Rydberg state [39]. The efficiency of laser ionization, ϵ_{laser}, is given by

$$\epsilon_{laser} = \frac{P_{ion}}{P_{ion} + P_{eff}}, \tag{4.2}$$

where $P_{ion,eff}$ is the ionization or effusion potential, respectively. The selectivity of this method over surface ionization is

$$\text{Selectivity} = \frac{\epsilon_{laser}}{\epsilon_{surface}}. \tag{4.3}$$

In the case of alkali metals, this is $\epsilon_{surface} > 5\,\%$ and $\epsilon_{laser} < 30\,\%$, hence the selectivity is once again dependent on the ionization potential of the element of interest [40].

4.2.3 *Isotope Separation*

The on-line mass separation of the isotopes of interest allows access to long chains of isotopes, making systematic studies possible. Two targets and adjoining separators are available at ISOLDE: the general purpose (GPS) and high resolution (HRS) separators. The GPS utilises one magnet to disperse the ions produced from its target. An electrostatic switchyard contains two pairs of electrostatic deflector plates, allowing three ion beams of neighbouring A/q to be selected. The movable plates allow two beams, either side of the central mass, to be selected, a maximum of ± 15 mass over charge units away. The lower- and higher-mass ion beams are then transported to the GLM and GHM beam lines respectively, enabling three experiments to be performed simultaneously [12].

The HRS is composed of two magnets and ion-optics to allow for a greater mass resolution of the ion beam, generally considered to be greater than 5,000 [2]. The two separators and corresponding beam lines are connected to the central beam line in the ISOLDE experimental hall by way of a 'merging switchyard' and cylindrical deflector plates. A result of the high voltage applied to the ions after ionization, an energy of up to 60 keV can be achieved. These ions are steered through the beam line with electrostatic ion optics to the necessary experimental set-up. The optics in the beam line consist of electrostatic elements, making the ion beam transport and steering through the line independent of mass [35].

The beam energy of 60 keV can be increased up to 3 MeV/A by charge-breeding and post-acceleration with REX-ISOLDE [41]. Improvements in the energy, intensity and quality of the ion beams produced at ISOLDE will be made with the development of HIE-ISOLDE in 2015, allowing up to 10 MeV/A to be eventually reached [42].

4.3 The ISCOOL Cooler-Buncher

Radioactive ions from HRS are cooled and trapped using the ISOLDE cooler-buncher, ISCOOL [43, 44]. This is a radio-frequency quadrupole (RFQ) cooler which traps the ions, decreases their velocity using a buffer gas, and releases them as a bunch with a 1–5 μs time focus. An example time-of-flight spectrum from ISCOOL to the end of the CRIS beam line is shown in Fig. 4.5. The average time-of-flight is 81.5 μs, with a time focus of 1 μs.

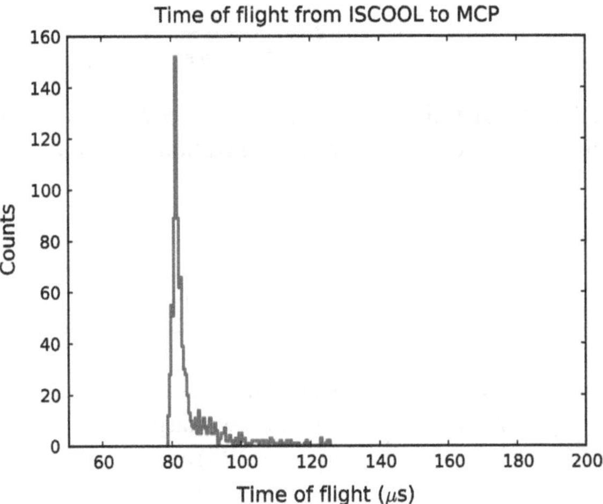

Fig. 4.5 The time taken for the ions, released from ISCOOL, to be detected on the MCP at the end of the CRIS beam line. The peak of the ions' arrival is 81.5 μs, with a full-width half-maximum of 1 μs. The tail of the time-of-flight spectrum is associated with space-charge saturation in ISCOOL

The bunched beam is re-accelerated to up to 60 kV using three extraction electrodes [45]. The cooling of the ion beam can be expressed in terms of emittance, with the direction of propagation through the cooler the z-direction. For an ion beam with waist Δx [mm] and divergence $\Delta\theta = \Delta p_x/p_0$ [mrad], the emittance in the x- or y-direction, the transverse emittance, ϵ_x, is given by

$$\epsilon_x = \Delta x \Delta\theta. \tag{4.4}$$

Similarly, the longitudinal emittance, ϵ_z [eV][μs], that in the z-direction, is regarded as being

$$\epsilon_z = \Delta p_z \Delta z, \tag{4.5}$$

where p_z and z have their usual definition of momentum and displacement in the z-direction. An application of the result by Liouville [46] demonstrates the emittance of an ion beam remains constant under the influence of conservative forces (such as those from an electromagnetic field). However, cooling of the ion beam involves non-conservative forces (through collisions with the buffer gas), with the addition of electrostatic forces for focusing and trapping the ions.

The RFQ cooler traps ions by confining them in an oscillating electric field. This RF electric field is produced by an electric quadrupole: four electrodes with opposite voltages applied to neighbouring electrodes. Two rods have a voltage of −(U+V $\cos\omega t$) applied, and two with +(U+V $\cos\omega t$). A rotating electric potential well is produced, confining the ions. A DC voltage is superimposed on the RF voltage by application to the segmented elements in the longitudinal direction, accelerating the

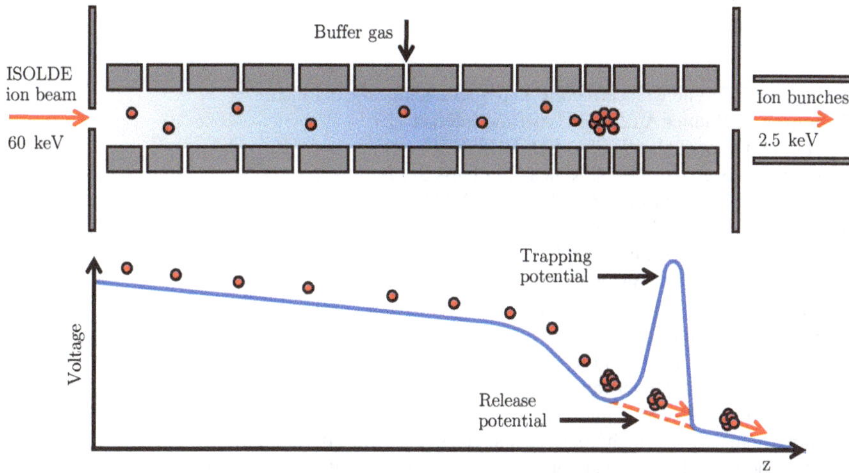

Fig. 4.6 A schematic of the ISCOOL cooler-buncher; (*Top*) The injection, cooling and extraction of the ion bunch. The oscillating electric field is produced by four (segmented) electrodes. Only two rods are shown in the figure for simplicity; (*Bottom*) The potential difference of the system, trapping the ions before releasing them as a bunch. The potential applied to the end plate is switched from 60 to 0 V to release the ion bunch

ions towards one end of ISCOOL. Figure 4.6 shows the DC voltage on the end plate of the cooler, guiding the ions towards the end of the cooler, trapping them in the process. The voltage on the end plate is switched from 60 to 0 V in less than 100 ns to release the ions. By adjusting the voltages in the trapping region, the bunch width can be modified. In addition, raising the voltage on the last few electrodes and setting the end-plate potential difference to zero, the time focus can be adjusted. This can be utilized to produce the optimum time focus of the ion bunch at the CRIS experimental set-up (although this technique was not used in the present work).

Effective cooling of the ion beam is present if $\text{mass}_{ion} > \text{mass}_{atom}$, with the minimum energy the ions can have being the thermal energy of the buffer gas itself. Due to space-charge constraints, the ions will always have a small energy distribution (they cannot be at zero velocity at the bottom of the electric potential well). The most efficiency cooling will occur with heavy ions and atoms comprising of a light buffer gas (typically helium). If however $\text{mass}_{ion} \approx \text{mass}_{atom}$ applies, then collisions in the RF field will cause heating, and no cooling will occur [47]. The energy distribution of the ions is also determined by the voltage ripple of the HV supply, corresponding to 0.001–0.1 V (on a voltage of between 30–60 kV).

The bunching of the ion beam with ISCOOL is essential for CRIS: pulsed lasers are used to step-wise excite the atoms, so overlapping the ion beam bunch with the laser pulse is more efficient since it reduces the duty cycle losses associated with a pulsed laser and a continuous ion beam [44].

References

1. http://te-dep-epc.web.cern.ch/te-dep-epc/machines/general.stm
2. Jonson B, Richter A (2000) Hyperfine Interact 129:1
3. Hansen PG et al (1969) Phys Lett B 28:415
4. Liberman S et al (1978) C R Acad Sci Paris Ser B 286, 353
5. Liberman S et al (1980) Phys Rev A 22:2732
6. Yagoda H (1932) Phys Rev 40:1017
7. Coc A et al (1985) Phys Lett B 163:66
8. Touchard F et al (1984) At Masses Fundam Constants 7:353–360
9. Bauche J et al (1986) J Phys B At Mol Opt Phys 19:L593
10. Duong HT et al (1987) EPL 3:175
11. Andreev SV et al (1987) Phys Rev Lett 59:1274
12. Kugler E (2000) Hyperfine Interact 129:23
13. http://home.web.cern.ch/about/experiments/ntof
14. http://home.web.cern.ch/about/accelerators/antiproton-decelerator
15. http://home.web.cern.ch/about/accelerators/cern-neutrinos-gran-sasso
16. http://home.web.cern.ch/about/accelerators/large-hadron-collider
17. Al-Khalili J, Roeckl E (2006) Euroschool Lect Phys Exotic Beams. Springer, Berlin
18. Al-Khalili J, Roeckl E (2004) Euroschool Lect Phys Exotic Beams. Springer, Berlin
19. Rare Isotope Science Assessment Committee (2007) National research council, scientific opportunities with a rare-isotope facility. The National Academies Press, US
20. Lettry J et al (1997) Nucl Instr Meth B 126:130
21. Kudryavtsev Y et al (2008) Nucl Instrum Methods Phys Res B 266:4368
22. Lassen J et al (2009) AIP Conf Proc 1104:9
23. Villari AC (2003) Nucl Instr Meth B 204:31
24. Äystö J et al (2012) Eur Phys J A 48:1
25. Cornell JC (2009) Final report of the EURISOL design study (2005–2009). GANIL, Caen, France
26. Sherrill B et al (1992) Nucl Instr Meth B 70:298
27. Kubo T et al (1990) Proceedings of the first international conference on radioactive nuclear beams, pp 563–572
28. Geissel H et al (1992) Nucl Inst Meth B 70:286
29. Müller AC (2000) EPAC 2000: 7th European particle accelerator conference p 730
30. Chouhan S et al (2013) IEEE Trans Appl Supercond 23:4101805
31. Okada K et al (2008) Phys Rev Lett 101:212502
32. Morrissey D et al (2003) Nucl Instr Meth B 204:90
33. Kubo T (2003) Nucl Instr Meth B 204:97
34. Geissel H et al (2003) Nucl Instr Meth B 204:71
35. Herlert A (2010) Nucl Phys News 20:5
36. http://isolde.web.cern.ch/isolde/
37. Kirchner R (1981) Nucl Instrum Methods Phys Res 186:275
38. Fedosseev VN et al (2012) Rev Sci Instrum 83(3):02A903
39. Mishin V et al (1993) Nucl Instrum Methods Phys Res B 73:550
40. Fedosseev VN et al (2012) Phys Scr 85:058104
41. Van Duppen P, Riisager K (2011) J Phys G Nucl Part Phys 38:024005
42. Lindroos M, Nilsson T (2006) Tech Rep CERN-2006-013, CERN, Geneva
43. Jokinen A et al (2003) Nucl Instr Meth B 204:86
44. Mané E et al (2009) Eur Phys J A 42:503
45. Aliseda IP et al (2004) Nucl Phys A 746:647
46. Goldstein H (1980) Classical Mechanics. Addison-Wesley, Boston
47. Nieminen KA (2002) Ph.D. Thesis, Department of Physics, University of Jyväskylä

Chapter 5
Collinear Resonance Ionization Spectroscopy

5.1 Laser Spectroscopy

Hyperfine structure and isotope shift studies are performed by measuring the frequencies and intensities of the atomic transitions between hyperfine states. When the energy of a photon equals the energy of the transition between two states, the photon excites the electron into this state. In the weak-field approximation, this electron then de-excites spontaneously and emits a photon (of the same frequency) which can be detected; or the electron can be further excited across the continuum and ionized, whereby the resonant ion is detected. Due to the hyperfine structure of the energy levels of the electrons, certain frequencies of laser light incident on the atom will give a resonance. An example hyperfine structure of the ^{178}Hf ground state and 8^- isomer can be seen in Fig. 5.1.

The centroid frequency (the centre of gravity of the hyperfine structure) is the frequency that corresponds to the energy difference between the two electronic states, when only the fine structure is accounted for. The isotope shift is the difference between centroid frequencies of two isotopes. This allows the mean-square charge radii of the second isotope to be calculated relative to the first, providing information on the changes in density, volume and shape of this isotope's nucleus, see Sect. 7.1.6.

Laser spectroscopy is employed in a variety of ways (e.g. collinear, in-source, resonance ionization) to exploit the advantages of each technique for the isotope under investigation [2, 3]. At the CRIS experiment, collinear laser spectroscopy and the more recent innovation of resonance ionization spectroscopy are combined, therefore the discussion of laser spectroscopy techniques will be limited to these methods.

5.1.1 Collinear Laser Spectroscopy

Collinear laser spectroscopy overlaps the laser with the atomic (or ion) beam in a collinear geometry. Doppler broadening associated with the thermal motion of the

K.M. Lynch, *Laser Assisted Nuclear Decay Spectroscopy*,
Springer Theses, DOI: 10.1007/978-3-319-07112-1_5

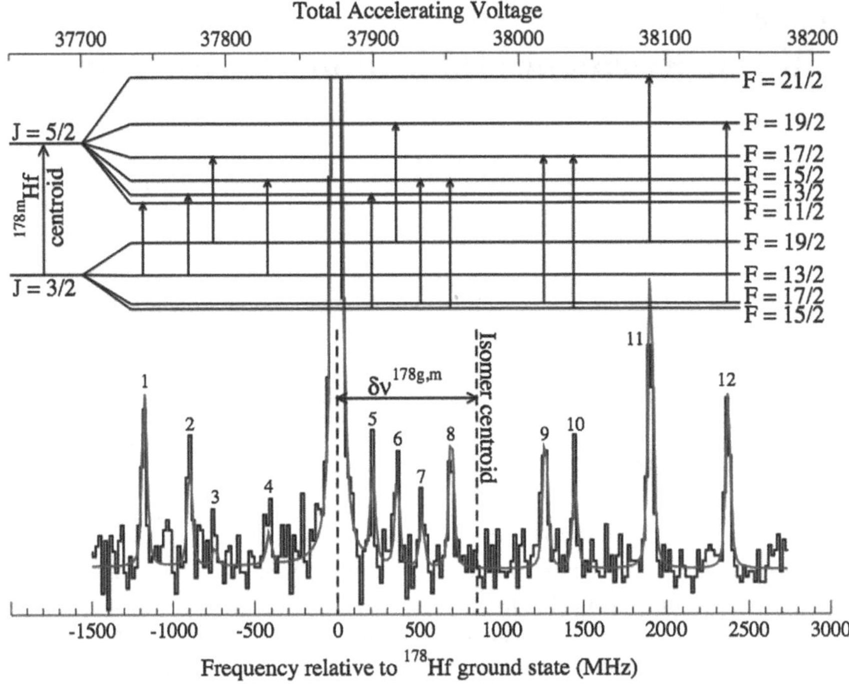

Fig. 5.1 The hyperfine structure of the ^{178}Hf ground state and 8^- isomer, using collinear laser spectroscopy on a bunched beam. Reproduced from Ref. [1]

atoms in the target-ion source (of the order of several GHz) is in general larger and can cover the full hyperfine structure. The collinear beams method gives a reduction in the thermal Doppler broadening by a factor of 10^3, to the MHz range. This reduction arises due to conservation of energy coupled to the kinetic energy definition which demands that the product of velocity and velocity spread remain constant under acceleration,

$$\Delta E = \delta(\frac{1}{2}mv^2) \approx mv\delta v,$$

which decreases the Doppler broadened line width of the hyperfine transition to below its natural width. An acceleration voltage of 40 kV will give a reduction in Doppler broadening comparable with the natural line-width of the hyperfine structure. This leads to a greater degree of selectivity between the ground and isomeric states of the nucleus, and a higher resolution and sensitivity of the nuclear observables. This is discussed in greater detail in Sect. 5.1.2.

A schematic of the collinear laser spectroscopy technique is shown in Fig. 5.2. A Doppler tuning voltage can be applied to the ions as they enter the charge exchange cell to be neutralized. This Doppler shifts the locked frequency of the lasers relative

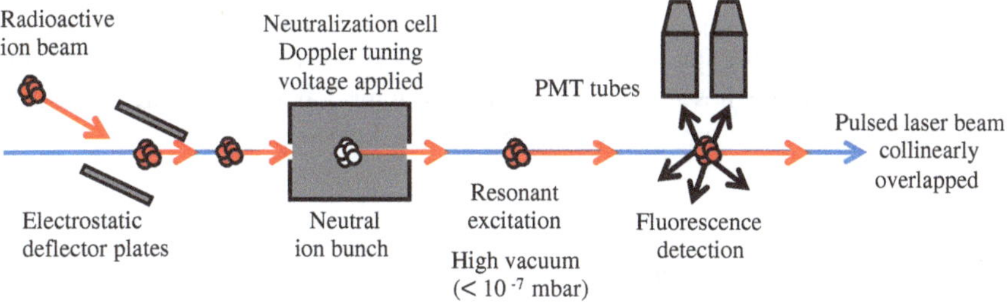

Fig. 5.2 A schematic of the collinear laser spectroscopy technique for atoms

to the interacting atomic bunch. The tuneable voltage is applied to the ions, tuning it on to resonance with the laser. The laser light, with a frequency ν_L in the laboratory frame, will have a Doppler shifted frequency ν when incident on the atom of

$$\nu = \nu_L \frac{\sqrt{1-\beta^2}}{1 \pm \beta}, \tag{5.1}$$

where β is defined as

$$\beta = \frac{v}{c} = \sqrt{\frac{2eU}{mc^2}}. \tag{5.2}$$

Here, v is the velocity of the atom (of mass m), c the velocity of light in vacuum, and U is the acceleration voltage. A $\pm$ sign in front of β indicates the direction of the laser beam with respect to the atoms (+ for a collinear direction, − for anti-collinear).

Resonance fluorescence is detected with a photo-multiplier tube (PMT) when the energy of the incident light matches the energy of a hyperfine transition. A detection efficiency of 1 % can be achieved with a continuous beam, but in general this technique is limited to at least 10^4 ions per second, due to the high background signal resulting from scattered laser light and PMT dark counts. Although in some cases hyperfine structure studies on yields of less than 10^3 ions per second are possible [4]. The distance from the region of resonant excitation to fluorescence detection (see Fig. 5.2) is minimized to avoid optical pumping into a 'dark state'. This occurs when the electron de-excites into an atomic energy level not probed by the laser light, resulting in this ion being lost to the excitation process. In order to minimise this loss, the spatial distance between the excitation region and the PMT tubes is made as small as possible.

The recent development of ion beam coolers has given rise to increased sensitivity in collinear laser spectroscopy due to the background suppression when using bunched beams. The technique was first established at the IGISOL facility in the University of Jyväskylä (Finland) [5] and introduced to great effect at ISOLDE in the form of ISCOOL [6, 7] (as discussed in Sect. 4.3) and at the ISAC facility at TRIUMF (Canada) [8]. The bunching of the beam reduces the background by the ratio of $t_{\text{bunching}}/t_{\text{bunch width}}$. With the use of ISCOOL to increase the sensitivity of collinear laser spectroscopy, a suppression of 10^4 is routinely reached. When the time between ion bunches $t_{\text{bunching}} \approx$ s and the temporal width of the ion bunch $t_{\text{bunch width}} \approx \mu$s, higher suppression can be achieved.

5.1.2 Resonance Ionization Spectroscopy

Resonance ionization spectroscopy (RIS) uses a similar technique, but ionizes the atomic bunch, and detects the ions incident on a particle detector, such as an micro-channel plate (MCP) instead of detecting the fluorescence photons with a PMT. This is achieved by step-wise exciting the atom across the ionization threshold and into the continuum. Resonant photons excite an electron in the atom from its ground state orbital to one or more excited states, and finally to an auto-ionizing state or directly in to the continuum. Field ionization from a high-lying Rydberg level is also possible.

From this method, two advantages over collinear laser spectroscopy arise. Firstly, the detection efficiency is greatly improved, as an MCP has close to 100 % detection efficiency. In contrast, a PMT has an efficiency of 3–20 % but the solid angle of detection for a PMT is approximately 10 % (at 10 cm) [9] compared to nearly 100 % solid angle coverage for an MCP. The second advantage comes from the ability to uniquely identify the presence of both the ground and isomeric states of an isotope, allowing for decay spectroscopy studies to be performed on purely isomeric beams.

Two conditions need to be satisfied in order for the ionization process to be considered saturated and the atoms ionized by the laser: the flux and fluence conditions. The flux condition states that the flux of photons, F, (ionizing the atom from the excited state) must be greater than the rate of depopulation, β, of the excited state into a dark state. With a cross section of ionization from the excited state σ_i, the flux condition is defined as

$$\sigma_i F > \beta. \tag{5.3}$$

The second criterion is the fluence condition, demanding that there are enough photons to ionize the atom

$$\frac{\sigma_i \psi g_2}{g_1 + g_2} \gg 1, \tag{5.4}$$

where $g_{1,2}$ are the statistical weights of the two states and ψ the fluence of photons. Satisfying these two criteria places demands on the type and power of the laser used. A typical value for $\sigma_i = 10^{-17}\,\text{cm}^2$ and $\beta = 10^7\,\text{s}^{-1}$, results in an flux of photons that must exceed $10^{23}\,\text{cm}^{-2}\,\text{s}^{-1}$. With a 5–20 ns pulsed laser, this is achievable with a power of 10 mJ/pulse, but is much harder when using a continuous wave laser.

5.1.3 Comparison with In-Source Laser Spectroscopy

The development of the RIS technique allowed the possibility of studying short-lived nuclei. In-source laser spectroscopy in both hot-cavity ion sources and gas catchers overlap multiple lasers within either a surface ionizer cavity (usually tungsten or tantalum) or an ion guide gas cell, ionising the atoms of interest. The gas cell technique has the ability to study superheavy elements (in the future facility of S^3 in GANIL, France), isomers and short-lived nuclei (at the IGISOL in Jyväskylä, Finland).

The selectivity of the hot-cavity in-source laser spectroscopy is limited by the Doppler broadening associated with the velocity of the ions in the surface ionizer cavity (at temperature of 2,500 K) and the laser bandwidth. Typical line width resolutions achieved with this technique range between 2 and 3 GHz. Similarly with in-gas-cell laser spectroscopy, the resolution of the technique is attributed to the laser line width and the pressure broadening, with the Doppler broadening contribution limited to room temperature (300 K) [10]. Additional reductions in sensitivity can occur from the plasma conditions that are a result of irradiating a gas with a laser. Such resolutions can be improved with the coupling of a gas jet apparatus (such as a Laser Ion Source Trap, LIST [10, 11]) to the gas cell catcher to reduce to Doppler broadening [12].

The selectivity limitations associated with in-source laser spectroscopy are most prevalent with lighter isotopes, making this technique suitable for isotopes with hyperfine structures of greater than 5 GHz (typically $Z < 60$). Collinear resonance ionization spectroscopy however combines the sensitivity of resonance ionization spectroscopy and the associated particle detection, with the resolution of collinear laser spectroscopy, reducing the pressure or Doppler broadening that occurs with in-source spectroscopy [2].

5.2 The CRIS Technique

The Collinear Resonance Ionization Spectroscopy (CRIS) experiment combines laser spectroscopy with decay spectroscopy: a unique technique that provides a wealth of information on the isotope under investigation [13, 14]. The combination of these two methods allows a beam heavily contaminated with radioactive isobars to be cleaned and decay spectroscopy performed on a pure sample. Additionally, it can also be used to identify the hyperfine components of an overlapping hyperfine structure associated with two nuclear states through the radioactive decay of its constituent parts.

The radioactive isotope of interest is produced at ISOLDE by impinging 1.4 GeV protons on a thick target (e.g. uranium carbide, UC_x). These radioactive products are then surface ionized, accelerated to 40 keV and mass separated with HRS to select the isotope of interest [15]. The production yields at ISOLDE for the neutron-deficient francium isotopes are presented in Table 5.1. The measured values are from an ISOLDE UC_x target with 0.6 GeV protons delivered from the

Table 5.1 Measured yield of francium isotopes per μC of integrated proton current

Isotope	Half-life	Yield (ions/μC)
202	0.34(4) s	7.1×10^1
203	0.55(2) s	1.0×10^3
205	3.9(1) s	1.7×10^5
206	15.9(2) s	2.4×10^6
207	14.8(1) s	3.6×10^6
211	3.10(2) min	1.5×10^8
218 m	1.0(6) ms	4.3×10^3
219	20(2) ms	8.9×10^3
220	27.4(3) s	3.8×10^7
221	4.9(2) min	2.8×10^7

The 0.6 GeV protons impinging on the UC_x target are delivered from the SC. The yield is expected to increase with 1.4 GeV protons delivered from the PS Booster

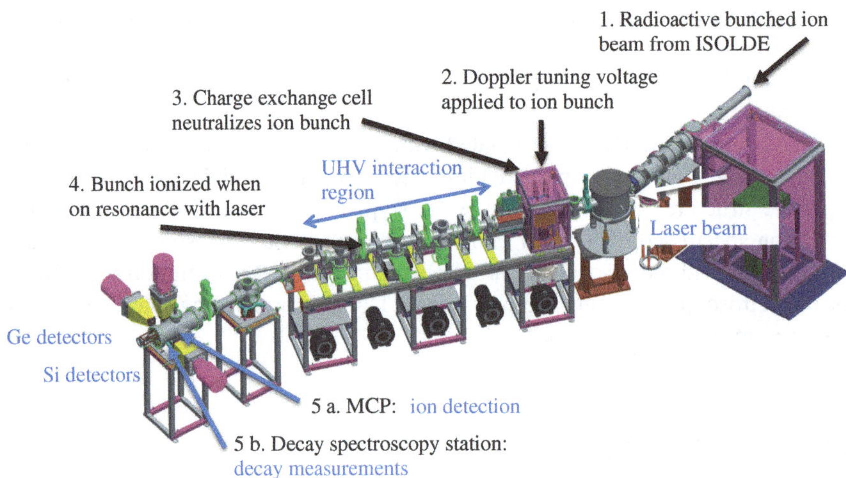

Fig. 5.3 The collinear resonance ionization spectroscopy (CRIS) beam line

synchro-cyclotron (SC). The yields of the francium isotopes since the upgrade from the SC to the PS Booster (1.4 GeV) have not been published but can be considered to be greater than those quoted in Table 5.1.

After selection of the francium isotope of interest, the ions are bunched using ISCOOL [7, 16]. The bunching of the ion beam is essential for removing the duty cycle losses due to the pulsed lasers. The ion beam is then deflected into the CRIS beam line, a 3D representation of which can be seen in Fig. 5.3.

After the ion beam is electrostatically deflected through a 30° bend, it is passed through an alkali vapour charge exchange cell (CEC). The CEC contains potassium metal, heated to a temperature of 150 °C, to form a vapour ($\sim 10^{-6}$ mbar). Due to the low ionization potential of the alkali metal potassium (4.34 eV [17]), charge transfer easily occurs. The positive francium ion acquires an electron, neutralising the ion bunch in preparation for resonance ionization.

In the interaction region, the atom bunch is collinearly overlapped with two laser beams. In this region ultra-high vacuum (UHV) is required, of the order of 10^{-9} mbar. This is to reduce the number of atoms that are non-resonantly ionized through collisions with gas molecules, which would result in a false count. To avoid contamination of the low vacuum in the interaction region with the alkali vapour from the CEC, differential pumping apertures are placed between the CEC and the UHV region. During the experimental campaign, the pressure in the interaction region reached 10^{-9} mbar. At this pressure, for a 0.1 pA beam at 30 Hz, the rate of non-resonant collision ionization would be almost negligible [18].

A sequence of two transitions at selected frequencies is required to bring the electron across the ionization potential, see Sect. 5.2.1. When the laser frequencies are on resonance with the hyperfine component of the optical transition, the isotope is ionized. By using short laser pulses, losses associated with optical pumping can be avoided. This process of resonance ionization [19] selects the isotope of interest. Additional selectivity is gained due to the kinematic shift between different masses of the isotopes caused by performing RIS on an accelerated beam. The sensitivity of the CRIS technique comes from the detection of resonant ions (with the possibility of 100 % detection efficiency), efficient ionization (between 10 and 30 %) and almost background free detection (from the low production rate of non-resonant ions).

After resonant ionization, the ions are electrostatically deflected through a 20° bend and detected with a micro-channel plate (MCP) housed in the decay spectroscopy station (DSS). The MCP detector (Model: Hamamatsu F4655-13) is set up as a negative ion counter. In this configuration, the beam strikes the surface of a copper plate and emits secondary electrons which are electrostatically guided to the MCP, resulting in the observed signal. More information on the set-up and constituent parts of the DSS can be found in Sect. 6.2.

Due to the high degree of selectivity and efficiency of the resonance ionization process, the CRIS technique can be utilized as a purification method [20]. This allows selective ionization of a particular state and decay spectroscopy on pure isomeric states to be performed. This technique of laser assisted nuclear decay spectroscopy is discussed in greater detail in Chap. 6.

5.2.1 Resonance Ionization of Francium

Figure 5.4 shows two of the possible resonance ionization schemes for francium. For the 'red' step, the atomic transition from $7s\,^2S_{1/2} \rightarrow 7p\,^2P_{3/2}$ has been experimentally determined by Coc [21] and the 'blue' step, from $7s\,^2S_{1/2} \rightarrow 8p\,^2P_{3/2}$, has been previously measured by Duong [22]. The $7p\,^2P_{3/2}$ state has a hyperfine structure almost three times larger than that of the $8p\,^2P_{3/2}$ state. The coupling of two 422.7 nm or 355 nm photons in the vacuum will non-resonantly ionize the atom,

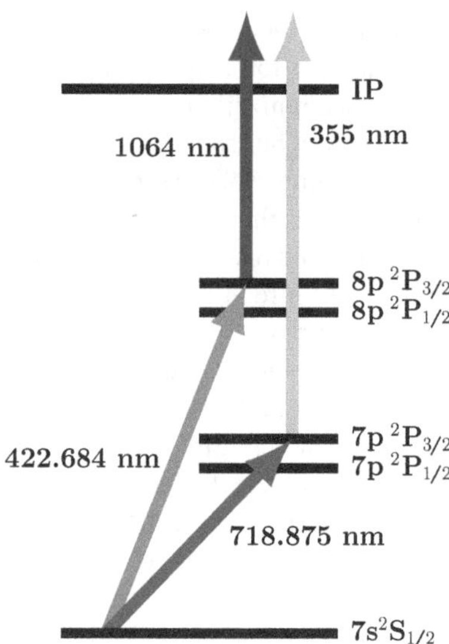

Fig. 5.4 The two possible two-step, resonance ionization schemes for francium. For this experimental campaign, the 422.7 nm+1,064 nm ionization scheme was used, probing the 7s $^2S_{1/2} \rightarrow$ 8p $^2P_{3/2}$ resonant transition

with both processes having similar cross-sections. Since the 422.7 nm laser light is however required for a resonant step, 10^6 times less laser power is required in comparison to the 355 nm step which is non-resonant. The rate of non-resonant ionization being less, it was decided to ionize the francium isotopes via the 8p $^2P_{3/2}$ state. These transitions are narrow enough to resolve the isotopes shifts (of the order of 10 GHz) and the hyperfine structure splitting (100 MHz). Scanning the frequency of the resonant (422.7 nm) excitation step probes the hyperfine structure resulting from the coupling between the 7s $^2S_{1/2}$ and 8p $^2P_{3/2}$ states with the nuclear spin.

For francium, the first transition is from the 7s $^2S_{1/2}$ electronic orbital to the 8p $^2P_{3/2}$ state with 422.7 nm light. During the August and October 2012 experimental runs, the 422.7 nm light was provided by the RILIS team at ISOLDE using a narrow-band Ti:Sa laser [23, 24] optically pumped by a diode-pumped solid-state (DPSS) laser (Model: DPSS 10 kHz) containing a Nd:YAG crystal. The fundamental frequency from the tuneable Ti:Sa laser (844 nm) was frequency doubled using a beta barium borate (BBO) crystal to produce the 422.7 nm laser light. This light was fibre-coupled into the CRIS beam line through 35 m of multimode optical fibre.

The second (non-resonant) transition is from the 8p $^2P_{3/2}$ state into the continuum using 1,064 nm light. This light was produced by a Nd:YAG laser (Model: Quanta-Ray LAB 130 30 Hz) next to the CRIS beam line. Both pulsed Nd:YAG lasers operated in Q-switch mode, where an optical switch waits 190 μs after the flash lamp fires to depopulate the inverted laser medium and produce the laser pulse

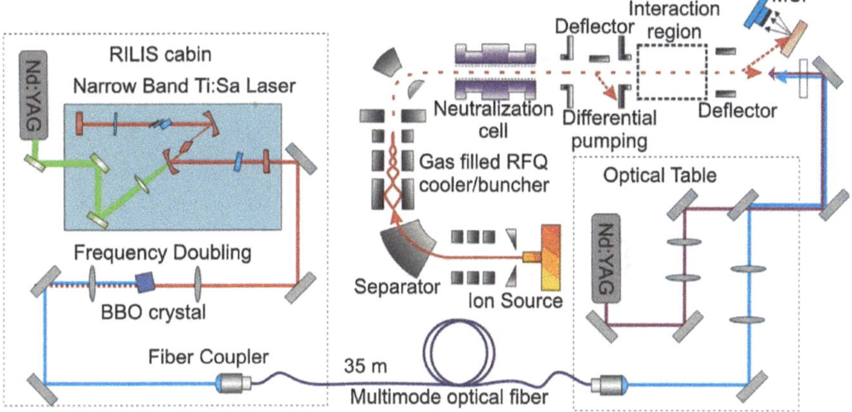

Fig. 5.5 A schematic of the lasers used during the CRIS experiment. The RILIS team provided the resonant 422.7 nm laser light, fibre-coupling it through 35 m of multimode optical fibre to the CRIS set up

of 20 ns. The 1,064 nm light was overlapped in time and space with the 422.7 nm laser beam and aligned through the beam line.

A schematic of the laser system is shown in Fig. 5.5. The atoms are ionized when the laser light is on resonance with an optical transition of the hyperfine structure and deflected to the decay spectroscopy station (DSS). The ions can be detected with an MCP (for hyperfine structure measurements) or implanted into a carbon foil for decay measurements, see Chap. 6 [25].

The lasers were aligned through the CRIS beam line with the use of a diode laser. This laser was installed at the end of the interaction region, where the non-ionizied beam is dumped. A glass window is present in this location, which transmits the laser light into the beam line. This allowed the optical mirrors to be installed in the correct location on the CRIS optical table next to the beam line for efficient transmission of the laser light. When the diode laser light was successfully aligned, minimal adjustment was required to send the 422.7 and 1,064 nm light originating from the CRIS optical table through the beam line.

5.2.2 Equipment Synchronization

For successful and efficient hyperfine structure measurements, the CRIS experiment relies upon the synchronization of the ion bunch with the two laser pulses in the interaction region, as shown in Fig. 5.6. In order to do this, a Quantum Composers eight channel pulse generator (Model: QC9258) is used to control the system. The master clock is the DPSS pump laser for the Ti:Sa laser, producing the 422.7 nm laser light at a rate of 10 kHz. This acts as the master trigger (a), triggering the pulse

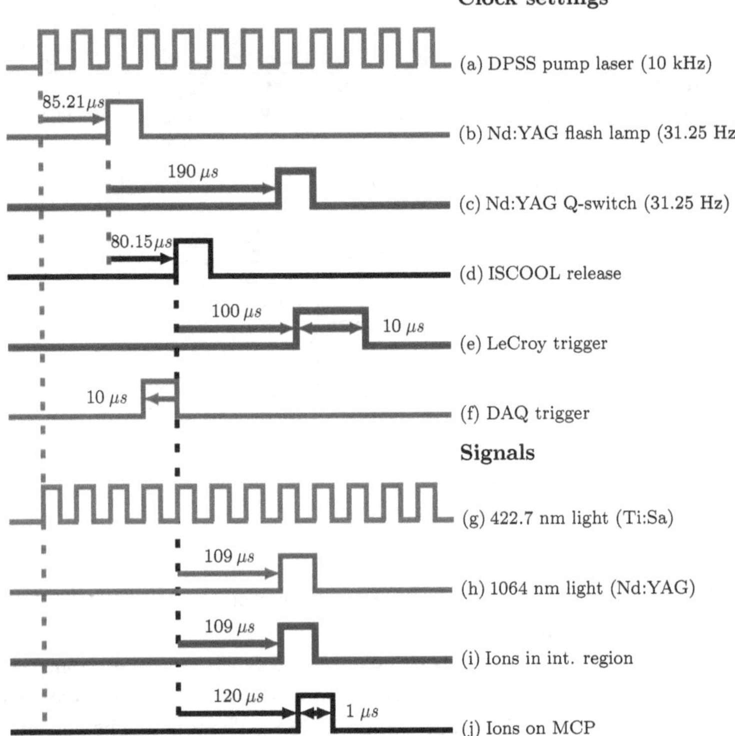

Fig. 5.6 The synchronization triggers outputted from the pulse generator repeated every 32 ms. [Clock settings] *a* The master trigger, from the DPSS laser at 10 kHz, triggered the pulse generator to output the following signals: *b* the trigger to the flashlamp of the Nd:YAG laser (which triggered the Q-switch *c* to fire the laser 190 μs later), *d* the trigger to ISCOOL to release the ion bunch, *e* the trigger to the oscilloscope to open a measurement window for data acquisition, and *f* the trigger to the data acquisition to save the ion counts. [Signals] The signals in the interaction region: *g* the 422.7 nm light, *h* the 1,064 nm light and *i* the ions arrive in the interaction region. *j* The resonant ions are detected by the MCP

generator to output an array of pulses. The first trigger (b) was sent to the flash lamp of the Nd:YAG, and after an internal delay of 190 μs, the Q-switch is triggered (c) and the laser fires. This laser has an adjustable repetition rate of between 27 and 33 Hz, so a frequency of 31.25 Hz was chosen to easily synchronize with the 10 kHz rate of the Ti:Sa laser. This meant that the trigger array repeated on a cycle of 32 ms, where 319 master trigger pulses were skipped. A second pulse (d) was sent to the ISCOOL cooler-buncher to set the potential on the end plate to zero and release the ion bunch. A third pulse (e) triggered the LeCoy oscilloscope which counted the number of ions impinging upon the MCP. This measurement window was open for 10 μs for each bunch. A final trigger (f) was sent to a computer to save the number of ions counted by the oscilloscope for each bunch. This signal was sent just before trigger

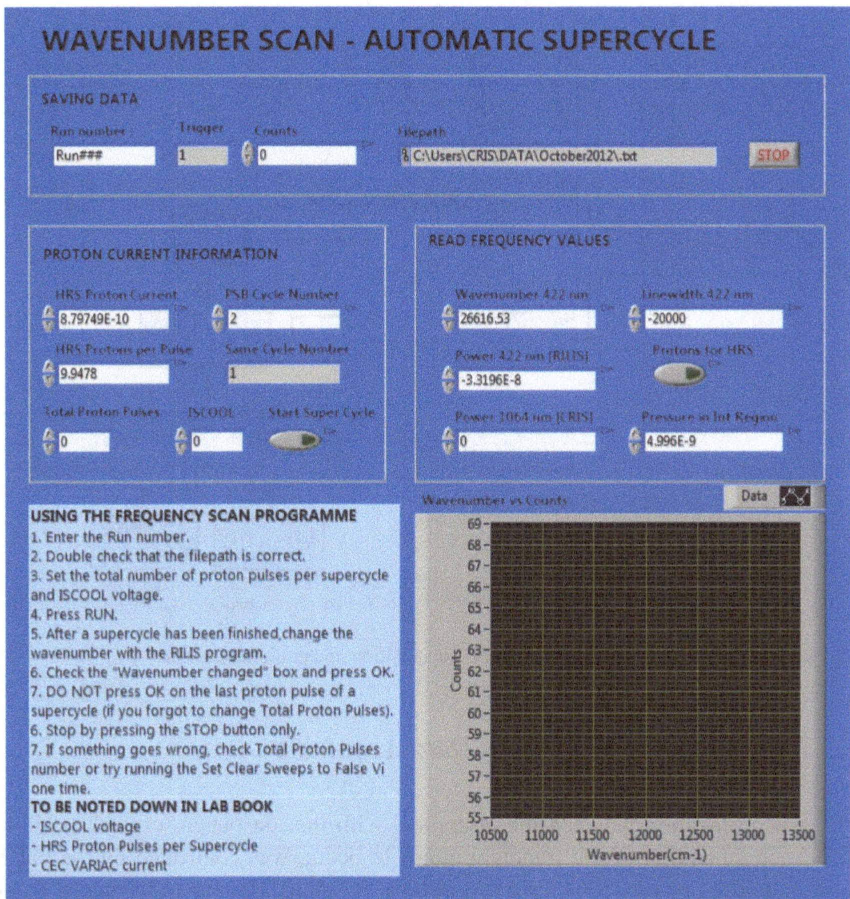

Fig. 5.7 Screenshot of the CRIS data acquisition and equipment control LabView programme

(d), releasing the ions from ISCOOL, in order to maximise the time the computer had to save the data before the next ion bunch arrived.

The arrival of the 422.7 nm light (from the Ti:Sa laser) and the 1,064 nm light (from the Nd:YAG laser) in the interaction region is shown in (g) and (h). The time of flight of the ions from ISCOOL to the MCP was measured to be 120 μs (i). The energy of the beam and the distance between the MCP and interaction region was used to calculate the time of flight of the ions from ISCOOL to the CRIS interaction region, which was 109 μs. More information on the dimensions of the CRIS beam line is given in Sect. 6.2.2. The arrival of the ion bunch is shown in (h).

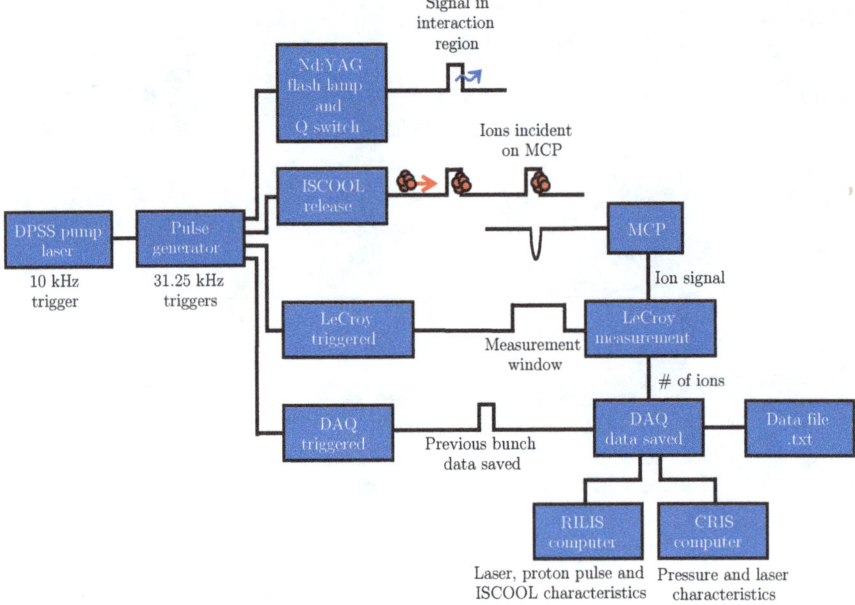

Fig. 5.8 Schematic diagram of the CRIS data acquisition and equipment control

5.2.3 Equipment Control and Data Acquisition

The synchronization between the lasers and with the ion bunch, critical to allow resonance ionization of the francium atoms to occur, was controlled by the pulse generator. The resonant ions were detected by the MCP. The electronic signal from the MCP was fully digitized with use of a LeCroy oscilloscope (Model: WavePro 725Zi 2 GHz bandwidth, 12 bit ADC, 20 GS/s), and when an ion was detected the oscilloscope registered the count. For each ion bunch, the oscilloscope counted the number of resonant ions and the number was saved by the CRIS data acquisition computer. The data acquisition system was written with LabView, National Instruments' graphical programming platform. A screenshot of the data acquisition programme is shown in Fig. 5.7.

The data acquisition programme was triggered by the pulse generator and saved a variety of experimental conditions. This was via a National Instruments PXI data acquisition board (Model: NI PXI-6225) connected to the computer via a USB interface. Once triggered, the programme saved the time, wavenumber and power of the two laser pulses, the pressure in the interaction region, the characteristics of the proton beam, the acceleration voltage of the ion beam, in addition to the number of ions detected in the ion bunch. A schematic of the data acquisition system is shown in Fig. 5.8.

The wavenumber and power of the 422.7 nm laser pulse, the characteristics of the proton beam and the acceleration voltage of the ion beam, was published as a shared variable by the RILIS data acquisition system and accessible through the technical network available at ISOLDE. The pressure in the interaction region and the power of the 1,064 nm laser light was measured by a separate computer at the CRIS set-up, and similarly published on the network. The number of ions counted by the oscilloscope was read by the data acquisition programme via an ethernet connection. The reading of the ion-number variable was the main source of delay in the data flow synchronization, and hence why a maximum time of 31 ms was introduced into the system to allow for any time lags.

The frequency of the resonant excitation step, the 422.7 nm laser light, was scanned to probe the 7s $^2S_{1/2} \rightarrow$ 8p $^2P_{3/2}$ atomic transition. The scanning and stabilization of the frequency was controlled by the RILIS Equipment Acquisition and Control Tool (REACT), a LabView control programme that allows for remote control, equipment monitoring and data acquisition [26]. This was achieved by controlling the tilt of the angle of the Ti:Sa laser system's etalon, adjusting the laser frequency. The francium experimental campaign at CRIS marked the first implementation of the REACT framework for external users. The remote control LabView interface ran locally at the CRIS setup, allowing independent laser scanning and control.

References

1. Bissell ML et al (2007) Phys Lett B 645:330
2. Cheal B, Flanagan KT (2010) J Phys G 37:113101
3. Blaum K et al (2013) Phys Scripta 2013:014017
4. Flanagan KT et al (2012) J Phys G Nucl Part Phys 39:125101
5. Nieminen A et al (2000) Hyperfine Interact 127:507
6. Jokinen A et al (2003) Nucl Instr Meth B 204:86
7. Mané E et al (2009) Eur Phys J A 42:503
8. Brunner T et al (2012) Nucl Instrum Methods Phys Res A 676:32
9. Kreim K, Ph.D. (2013) Thesis, Max Planck Institute for Nuclear Physics, University of Heidelberg
10. Sonoda T et al (2009) Nucl Instrum Methods Phys Res B 267:2918
11. Reponen M et al (2011) Nucl Instrum Methods Phys Res A 635:24
12. Kudryavtsev Y et al (2013) Nucl Instrum Methods Phys Res B 297:7
13. Procter TJ et al (2012) J Phys Conf Ser 381:012070
14. Lynch KM et al (2012) J Phys Conf Ser 381:012128
15. Jonson B, Richter A (2000) Hyperfine Interact 129:1
16. Frånberg H et al (2008) Nucl Instrum Methods Phys Res B 266:4502
17. http://www.webelements.com
18. Procter T, Flanagan K (2013) Hyperfine Interact 216:89
19. Billowes J, Campbell P (1995) J Phys G Nucl Part Phys 21:707
20. Letokhov V (1973) Opt Commun 7:59
21. Coc A et al (1985) Phys Lett B 163:66
22. Duong HT et al (1987) EPL 3:175
23. Rothe S et al (2011) J Phys Conf Ser 312:052020
24. Fedosseev VN et al (2012) Rev Sci Instrum 83(3):02A903
25. Rajabali MM et al (2013) Nucl Instrum Methods Phys Res A 707:35
26. Rossel R et al (2013) Nucl Instrum Methods Phys Res B 317, Part B, 557

Chapter 6
Laser Assisted Nuclear Decay Spectroscopy

The novel technique of laser assisted nuclear decay spectroscopy was developed at the CRIS beam line to take advantage of the ultra-pure ion beams produced by the resonance ionization method. In addition to radioactive studies of pure ground state or isomeric beams, the decay spectroscopy setup can be used to alpha-tag the hyperfine components of overlapping structures. This allows the hyperfine structure of two states to be separated by exploiting their characteristic radioactive decay mechanisms. This results in a smaller error associated on the hyperfine parameters, and hence a better determination of the extracted nuclear observables.

6.1 Production of Isomeric Beams with CRIS

Laser radiation is used to resonantly excite and ionize an atomic beam, probing the hyperfine structure of an isotope. From the frequency at which these atomic transitions occur, nuclear obervables (such as isotope shift, mean-square charge radii, magnetic dipole and electric quadrupole moments) can be extracted. In the experimental campaign to measure the hyperfine structure of francium, two transitions at selected frequencies are required to bring the electron across the ionization potential. When the laser frequency is on resonance with the hyperfine component of the atomic transition, the isotope is ionized. This process of resonance ionization [1, 2] selects the isotope of interest, see Chap. 5.

The HRS and GPS magnets offer a great degree of mass separation (A/q), but are not sensitive enough to the change in mass of isomeric states. The maximum selectivity from resonance ionization of an isotope is given by

$$S = \prod_{n=1}^{N} \left(\frac{\Delta\omega_{AB,n}}{\Gamma_n}\right)^2 = \prod_{n=1}^{N} S_n, \tag{6.1}$$

where $\Delta\omega_{AB}$ is the separation in frequency of the two states, Γ_n is the FWHM of the state, S_n is the selectivity of the transition and N is the number of transitions used.

K.M. Lynch, *Laser Assisted Nuclear Decay Spectroscopy*,
Springer Theses, DOI: 10.1007/978-3-319-07112-1_6

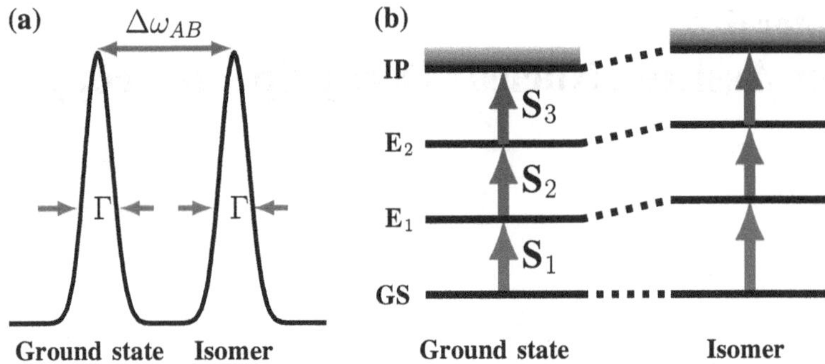

Fig. 6.1 **a** The difference in frequency $\Delta\omega_{AB,n}$ between two absorption lines, due to the isomer shift. **b** The number of resonant transitions increases the selectivity of the process

Figure 6.1a shows this schematically. Additional selectivity can be gained from the kinematic shift between different masses of the isotopes since the laser is overlapped with an accelerated beam. The total selectivity of a resonance ionization process is given by the product of the individual selectivities, thus the higher the number of atomic excitation steps, the greater the selectivity of the nuclear state, illustrated in Fig. 6.1b. Indeed, the selectivity of resonance ionization can be compared to the mass resolution of a state in mass spectrometry. The sensitivity of the technique comes from the detection of resonant ions (with the possibility of 100 % detection efficiency), efficient ionization (10–30 %) and almost background free detection.

Due to the high selectivity and efficiency of the resonance ionization process, the CRIS technique can be utilized as a purification method [3]. The different nuclear observables (spin, moments, mean-square charge radii) of isotopes and isomers produce different hyperfine structures, resulting in the lasers only ionizing the nuclear state of interest while on resonance with a characteristic transition.

For ^{221}Fr, the resulting selectivity can be naively calculated to be $(20/1.5)^2$ per resonant transition but in reality, the selectivity is higher. Equation 6.1 holds for Lorentzian absorption line profiles, but the profile of the laser used during the experimental campaign was Gaussian (1.5 GHz line width). The tail of the Gaussian profile will decrease at a faster rate relative to a Lorentzian profile, resulting in the Lorentzian component of the 1.5 GHz absorption line being much smaller. Hence in principle, decay spectroscopy can be performed on pure isomeric states with a suppression of the ground state by a factor of at least 10^4 per transition.

6.2 The Decay Spectroscopy Station

The decay spectroscopy station (DSS) consists of a rotatable wheel implantation system, shown in Figs. 6.2 and 6.3. It is based on the design from KU Leuven [4] (Fig. 1 of Ref. [5]), which has provided results in a number of successful

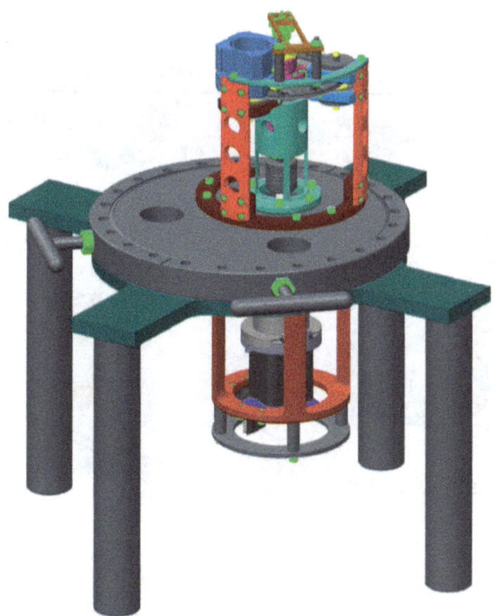

Fig. 6.2 A 3D representation of the steel wheel and stepper motor for controlling the rotation of the wheel

experiments [5–7]. The wheel holds 9 carbon foils, produced at the GSI target laboratory [8], with a thickness of 20(1) μg cm^{-2} (90 nm) into which the ion beam is implanted. The carbon foils and wheel system are favoured over a Mylar tape system for the compact size and, more importantly, better detection efficiency. This is a result of the improved transmission of alpha particles following the decay of implanted ions in the carbon foils.

A stepper motor (Model: 17HS-240E, 2.3 A/phase) can be used to rotate the wheel. The controller for the motor is operated via a programmable computer software package. The rotation of the wheel can be controlled using optical switches or by predefining the number of steps that the motor should turn. In order to maintain UHV, the motor is situated externally and connected to the internal wheel system via a magnetic coupling supplied by UHV Design. This coupling has negligible influence on the noise introduced to the signal between the detector and the preamplifier.

During the experimental campaigns in August and October 2012, the wheel was manually rotated as required. This was due to the low rate of revolutions required in this experiment: implantation into one carbon foil took place every 5–20 minutes when laser assisted nuclear decay spectroscopy was performed. The highest rate of rotation of the steel wheel was changing to the next carbon foil every minute, easily achievable by hand. Figure 6.2 shows the vacuum flange (on a stand) that the stepper motor and the steel wheel is mounted on. The flange is installed onto the far-side of the DSS chamber. Removal of the flange allows easy installation of the detectors and carbon foils.

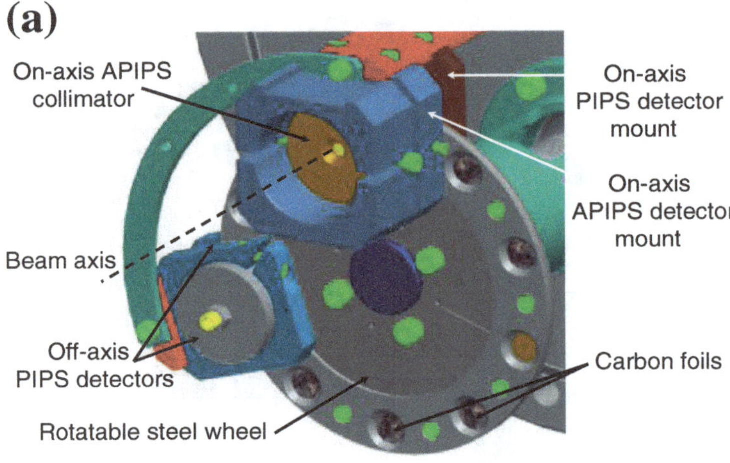

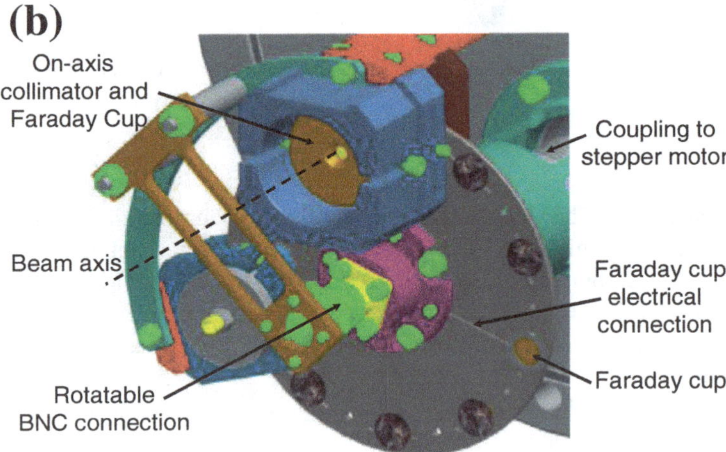

Fig. 6.3 **a** A 3D representation of the rotatable wheel and silicon detectors of the decay spectroscopy station 'windmill' system. **b** The same windmill system is shown with the carbon-foil ring-holder removed in order to highlight the Faraday cup

6.2.1 Alpha-Particle and Gamma-Ray Detection

Two Canberra Passivated Implanted Planar Silicon (PIPS) detectors for charged-particle detection (e.g. alpha, electron, fission fragments) are situated on either side of the implantation carbon foil, as shown in Fig. 6.3a. One PIPS detector (Model: BKA 300-17 AM, thickness 300 μm) sits behind the carbon foil and another annular PIPS (APIPS) (Model: BKANPD 300-18 RM, thickness 300 μm, with an aperture of 4 mm) is placed in front of the carbon foil. The detectors are connected to charge sensitive Canberra preamplifiers (Model: 2003BT) via custom-made Kapton cables and a UHV type-C sub-miniature electrical feed-through. The ion beam is directed through the 4 mm aperture of the APIPS detector and implanted into the carbon foil. Decay products from the carbon foil can be measured in either the PIPS or APIPS detectors. In addition to the on-axis detectors, two PIPS detectors can be placed off-axis, located at 108° away from the on-axis pair. Rotating the implanted carbon foil from the on-axis to off-axis detector position allows the measurement of unperturbed decay curves for longer-lived decay products.

For gamma-ray detection, up to three high-purity germanium (HPGe) detectors can be placed around the implantation site. During the August and October 2012 experimental campaigns, two germanium detectors were used: a E-ΔE germanium detector (Model: GR3019/GL1512 telescope) and a 70 % HPGe detector (Model: GC7020). These were placed a right-angles to, and above, the implantation site of the carbon foils, respectively.

6.2.2 Beam Tuning to the Carbon Foil

Resonantly ionized bunched beams from the interaction region in the CRIS beam line are deflected to the DSS by applying voltage to a pair of vertical electrostatic plates. The deflected ion beam is transported along the length of a 1.93 m long beam line before implantation into a carbon foil in the DSS. This dimension, in addition to the extra 60.0 cm to the centre of the interaction region, was used to calculate the time of flight of the ions from the interaction to the MCP. During the ion flight from the 20° bend to the DSS, the beam has to be directed through a 64.9 mm spacing between the copper plate and MCP (see Fig. 6.4) in addition to being focused through the 4 mm aperture of the APIPS detector. To achieve this, ion optics components have been added to the beam line as shown in Fig. 6.4. After the 20° deflection, the beam is first directed through a quadrupole doublet which corrects for astigmatism introduced earlier in the beam line. It is then passed through a final pair of vertical and horizontal steering plates (see Fig. 6.4) which, along with the quadrupole doublet, allow the focus and position of the beam to be optimized for either the copper plate or the carbon foil.

Two sets of collimators are installed in the detection chamber. The first is attached to a manual linear actuator with a series of 3, 6 and 10 mm apertures and is housed in the same region as the MCP detector. The second is a fixed aluminium disc placed in front of the APIPS detector to protect against direct ion implantation into the silicon wafer, see Fig. 6.3b. The PEEK[1] holder electrically insulates the disc from the grounded support structure as well as the APIPS detector. An electrical contact is made to the collimator, allowing the current generated by the ion beam when it strikes the collimator to be measured and the plate to be used as a beam monitoring device. When it is not in use, it is electrically grounded to avoid charge build-up.

[1] Polyether ether ketone.

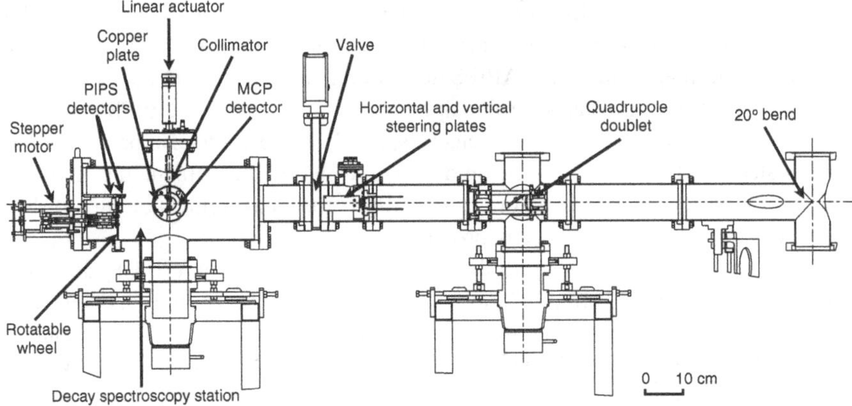

Fig. 6.4 Scaled drawing of the CRIS beam line after the interaction region. The ion optics components used to guide the beam to the decay station are indicated

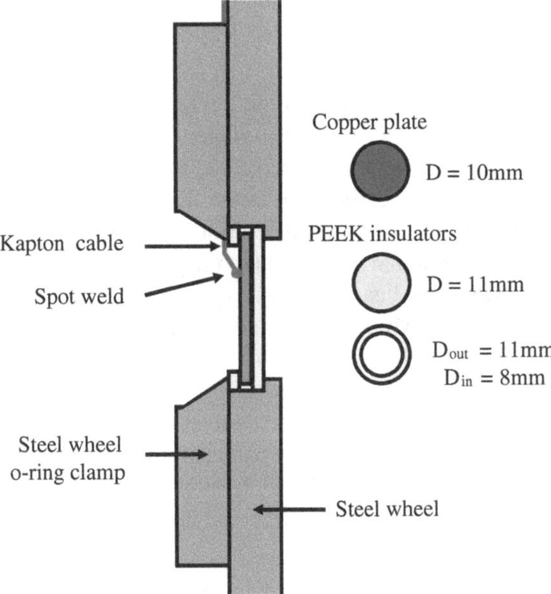

Fig. 6.5 A schematic of the Faraday cup installed in the DSS

During the first experimental run in November 2011, beam tuning to the carbon foil was carried out with a radioactive ion beam. The APIPS detector was used to monitor the signals and resulted in considerable contamination of the collimator.

Consequently, a Faraday cup was designed and installed in the location of one of the carbon foils, shown in Fig. 6.3b. This plate (thickness 0.5 mm, diameter 10 mm) is electrically isolated from the steel wheel by PEEK rings and connected by a spot-welded Kapton cable attached to a rotatable BNC connection in the centre of the wheel, shown in Fig. 6.5. These additions increased the total transmission efficiency of the CRIS beam line from less than 10 % to over 35 %.

6.2.3 Data Acquisition

The DSS data is acquired with a fully digital data acquisition system (DAQ), consisting of XIA digital gamma finder (DGF) revision D modules [9]. Each module has four input channels with a sampling rate of 40 MHz. Signals from the silicon and germanium detector preamplifiers are fed into separate modules on account of the different decimation needed to read in the pulses of different temporal length. Decimation in these DGF modules corresponds to an averaging of sampling points (25 ns apart) in increments of multiples of 2 [9]. Signals from the silicon detector preamplifiers are faster than those from their germanium counterparts, therefore decimation times of 2 (100 ns) and 4 (400 ns) are used, respectively.

A schematic diagram of the DSS data acquisition system can be seen in Fig. 6.6. All signals fed into the digital DAQ are self-triggered with no implementation of master triggers. In addition to the detector signals, the logic pulses from the release of the proton pulse and the release from the ISCOOL cooler-buncher can also be fed into the digital DAQ. The logic signals can then be used during offline analysis for correlating the arrival of the ion bunches in the DSS to the decay events measured by the silicon detectors. These signals however were not implemented during August and October 2012 for simplicity in the first experiments. The DGF DAQ is a modular unit: the number of channels can be increased by adding DGF modules to the DAQ system.

Due to the reflective surface of the inside of the vacuum chambers, when the 1,064 nm laser light was delivered through the CRIS beam line, it reflected along the chamber walls. Despite the collimator in front of the APIPS detector to protect it from ion implantation (and laser light), the infra-red light caused a shift in the baseline of the signal from the silicon detector. This required that the parameters for the DGF modules be adjusted to account for this effect online, since the reflections were due to the particular setup of the experiment.

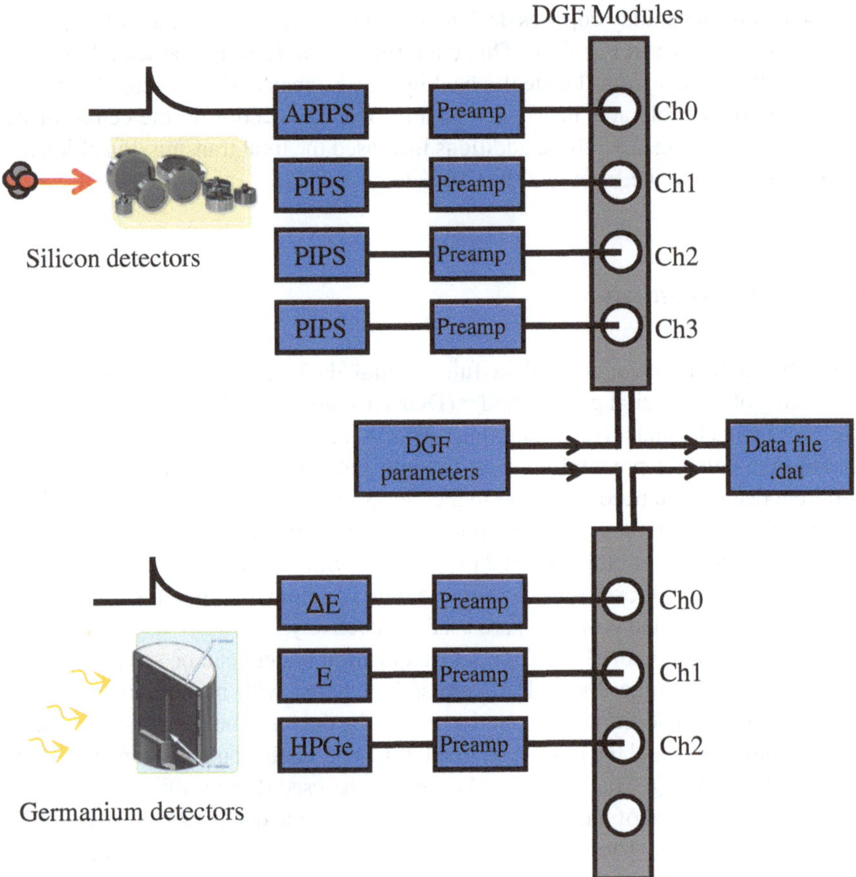

Fig. 6.6 Schematic diagram of the DSS data acquisition system

6.3 Characterising the DSS

6.3.1 Inherent Shielding in the DSS

Measurements of the gamma-ray attenuation by the stainless steel body of the chamber and the steel windmill were performed using a 70 % HPGe detector (Model: GC7020). The results are presented in Figs. 6.7 and 6.8, respectively.

When measuring the attenuation of the steel wall of the DSS chamber, known gamma-ray sources of ^{60}Co, ^{152}Eu and ^{241}Am were positioned 65 mm from the detector, with and without the chamber present. More than 92 % of the 59 keV gamma rays and 52 % of the 123 keV gamma rays were attenuated (see Fig. 6.7) by the 3 mm thick steel walls. At energies above 591 keV, fewer than 18 % of the gamma rays were affected by the steel wall. This high attenuation at low energies is due to the high Z

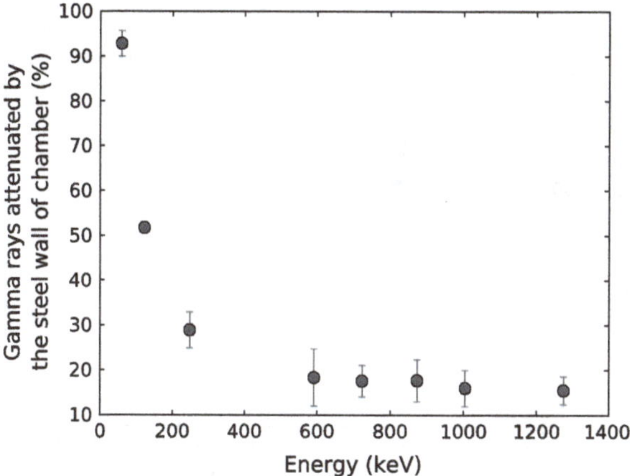

Fig. 6.7 The percentage of gamma rays attenuated by the steel wall of the DSS chamber. More than 92 % of gamma rays of less than 60 keV are attenuated. At energies above 590 keV, less than 18 % of gamma rays are stopped by the steel wall of the chamber

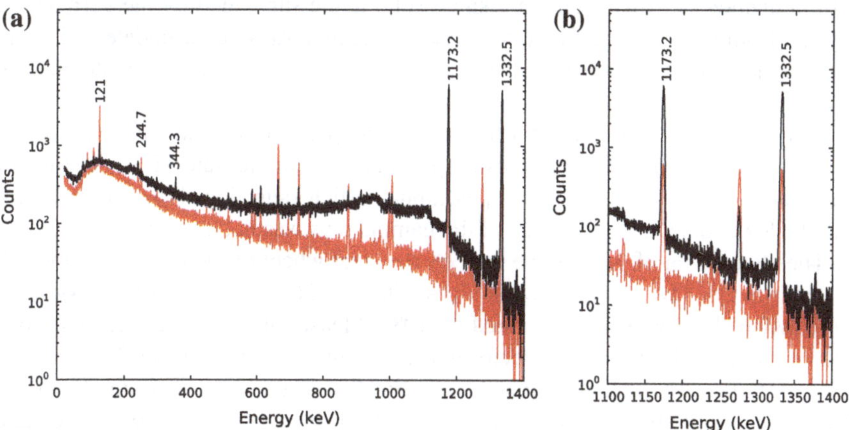

Fig. 6.8 Attenuation of gamma-rays by the steel wheel of the DSS. The *black* spectrum shows the ^{60}Co source in front of the germanium detector and the ^{152}Eu source 108° away from the detector. The *red* spectrum shows the reverse

of the stainless steel material of the DSS chamber walls. Despite the walls being as thin as possible for manufacturing (3 mm), they still provide a significant attenuation to the low-energy gamma rays. For this reason, aluminium chambers are usually favoured over steel, but would not give the required vacuum conditions necessary for the CRIS experiment. Enhanced vacuum was favoured over gamma-ray detection efficiency, and the DSS chamber was manufactured out of steel. Additional reduction in detection efficiency is related to the position of the germanium detectors in relation to the implantation site. As can be seen in Fig. 6.4, the location of the germanium

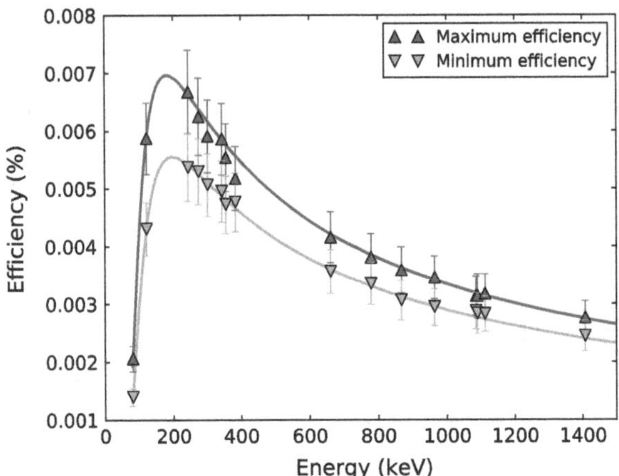

Fig. 6.9 The absolute photo-peak gamma-ray detection efficiency of the germanium detector

detectors are far away from the carbon foil due to the dimensions of the chamber (20 cm diameter). This results in a small solid angle subtended by the germanium crystal, leading to a large fraction of emitted gamma rays being undetected. Both these issues are addressed in the new design of the DSS chamber, as detailed in Sect. 6.4.

However, it must be noted that the primary purpose of the DSS was to alpha-tag the overlapping hyperfine structures of ground and isomeric states. For the study of the neutron-deficient francium isotopes, alpha-particle detection alone was sufficient and the lower detection efficiency of the gamma rays was not a priority.

The attenuation of the gamma rays by the steel wheel of the rotatable windmill system was also investigated. Two sources, ^{60}Co (2.156 kBq) and ^{152}Eu (1.249 kBq), were attached to the wheel separated by 108°. Measurements were taken with one source placed in front of the detector while the other was 108° away. The gamma-ray energy spectra obtained are shown in Fig. 6.8. The black spectrum shows the ^{60}Co source directly in front of the germanium detector and the ^{152}Eu source at a position 108° furthest away from the germanium detector. The red spectrum shows the ^{152}Eu source in front of the detector and ^{60}Co in the 108° position. From Fig. 6.8b, the characteristic ^{60}Co lines are decreased by a factor of 10 at 1.3 MeV when the ^{60}Co source is rotated away from the germanium detector. The change in solid angle coverage between the two positions accounts for approximately 70 % of this reduction. The remaining 30 % is the result of gamma-ray attenuation by the steel wheel. This provides a shielding advantage for the implanted ions in an off-axis foil when studying decays at the on-beam axis station and vice versa.

6.3.2 Efficiency of the Germanium Detectors

A 90 % HPGe detector (Model: GC9019-7935-7) was placed beside the DSS chamber at the on-beam axis location. The 90 % relative efficiency of the detector is relative to a 3″ × 3″ NaI crystal at 1.33 MeV. The absolute efficiency of this detector was measured for gamma rays emitted at the location of the carbon foil using three sources: ^{133}Ba, ^{137}Cs and ^{152}Eu.

The gamma-ray sources could not be placed in the exact position of the foil due to spatial constraints. Instead, the sources were attached individually to both the near-side and far-side locations of each detector. This allowed a maximum (near-side) and minimum (far-side) detection efficiency to be calculated for each gamma-ray energy, as shown in Fig. 6.9. The presence of the APIPS and PIPS detectors was also accounted for in the efficiency measurement. A maximum efficiency of 0.33 % and a minimum efficiency of 0.29 % was determined at 1 MeV.

6.3.3 Solid Angle Coverage of the Silicon Detectors

The PIPS detector was placed at a distance of 1 mm from the carbon foil to maximize the solid angle coverage of 45 %, see Fig. 6.10 (Blue line). The APIPS detector was positioned to optimize the solid angle of detection while minimizing the solid-angle loss through the aperture of 4 mm. The pure solid angle subtended by the APIPS detector, shown in Fig. 6.10 (Red line) was calculated from a point source located in the centre of the carbon foil with no hindrance. A distance of 4 mm from the carbon foil was chosen to maximise the solid angle coverage of the APIPS detector of 17 %.

6.3.4 Calibration of the DSS

Figures 6.11 and 6.12 show the calibration curves for the APIPS silicon detector, and the two germanium detectors, Ge0 and Ge2, respectively. After installation in the DSS, three of the four silicon PIPS detectors did not respond to biasing. Rather than risk damaging the silicon detectors by applying voltage to them, only the annular APIPS detector was used in the experiment. The results are reported in Chap. 7. This silicon detector was calibrated using the known alpha-particle energies from the decay of ^{221}Fr. Implantation of the reference francium isotope ^{221}Fr into a carbon foil was made after optimization of the ion beam tuning with ^{238}U. The alpha-particle energies from the decay of both ^{221}Fr and its daughter (and grand-daughter) isotopes ^{221}Ra, ^{217}At, ^{217}Rn and ^{213}Po were used to produce a calibration curve, see Fig. 6.11, at the appropriate energy range for the francium isotopes. In addition to a linear fit to the data points, a quadratic fit was investigated. This gave a reduced χ^2 goodness-of-fit of 10.1, compared to a χ^2 of 14.1 for the linear fit. Due to the

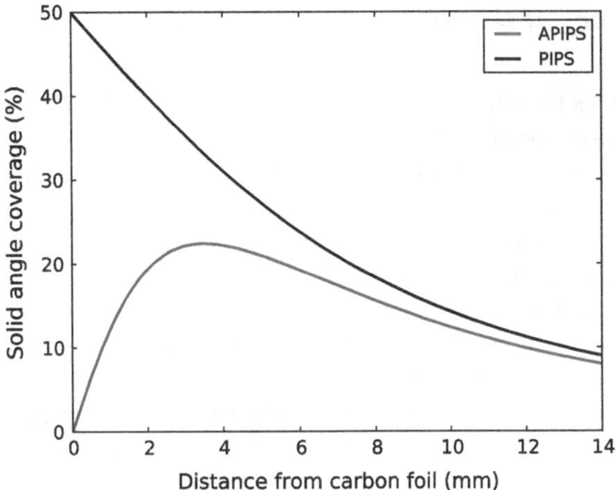

Fig. 6.10 The solid angle coverage of the PIPS (*blue*) and APIPS (*red*) silicon detectors, placed at 1 mm and 4 mm from the carbon foil, respectively, maximized the solid angle coverage. For Color interpretation see online version

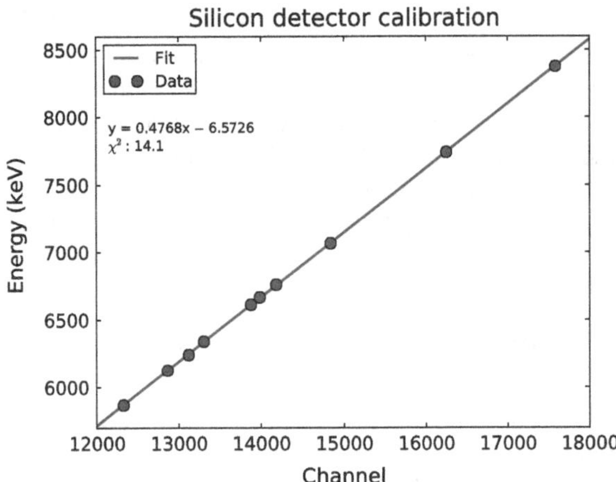

Fig. 6.11 Calibration curve for the APIPS detector

minimal improvement in the fitting procedure with a quadratic fit, the simpler linear equation was used to calibrate the APIPS detector.

During the experiment, the thick crystal (Ge1) of the E-ΔE germanium detector also stopped responding to biasing. This meant that only the ΔE planar crystal (Ge0) of the detector was able to detect gamma rays, in addition to the 70 % HPGe detector (Ge2). A calibration source of ^{152}Eu (1.249 kBq) was placed on the DSS, next to each of the germanium detectors in turn, and a measurement was taken. Figure 6.12

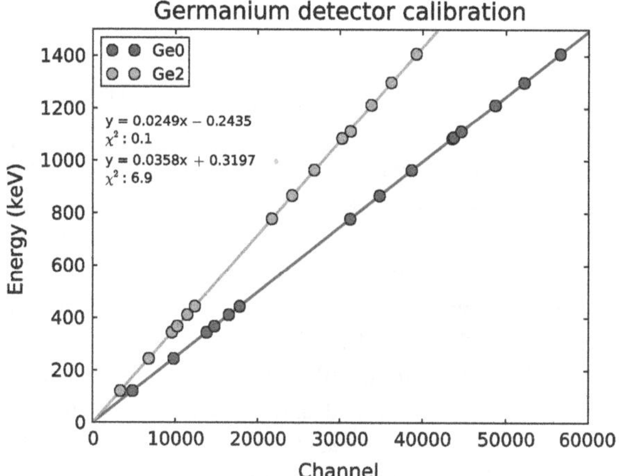

Fig. 6.12 Calibration curve for the germanium detectors

shows the calibration curve for both the Ge0 and Ge2 detectors (the ΔE and the HPGe detectors) used to calibrate the gamma-ray energy spectra.

6.3.5 Energy Resolution of the Silicon Detectors

The energy resolution (FWHM) of the silicon detectors was investigated off-line using a quadruple alpha source (^{148}Gd, ^{239}Pu, ^{241}Am, ^{244}Cm) placed in the position of a carbon foil. For comparison, both an analogue DAQ system as well as the digital DAQ were used for this measurement. The results for both systems were consistent with a FWHM value of 18 keV (at 5.485 MeV from the alpha decay of ^{241}Am) for the PIPS detector and 19 keV for the APIPS detector. These values agree favourably with the manufacturer's specifications of 17 keV for the PIPS and 18 keV for the APIPS detector.

During the experimental run, the energy resolution of the APIPS detector was not as high. This was due to the difficulty experienced when setting up the digital DAQ on-line to account for the change in baseline (see Sect. 6.2.3): the DGF parameters had to be optimized with a radioactive beam of ^{221}Fr with a half-life of 4.9 min. This isotope was implanted into the carbon foil for 10 s at a time and the DGF parameters optimized for the signals seen by the APIPS detector. The decreasing number of alpha particles emitted from ^{221}Fr, compared to the constant rate of alpha particles emitted from the quadruple alpha source (these alpha sources had much longer half-lives) meant that optimising the DGF parameters for high alpha-energy resolution was challenging. For this reason, an energy resolution of only 30 keV at 6.341 MeV was achieved.

6.4 The Upgrade of the DSS

Due to the low gamma-ray detection efficiency in the current DSS, work has begun on a new chamber design in order to reduce the attenuation of gamma rays and increase the solid angle subtended by the germanium detectors. This new design can be seen in Figs. 6.13 and 6.14, showing isometric and side view projections.

The new apparatus will be attached to the original DSS, downstream of the CRIS beam line, replacing the flange with the windmill and stepper motor attached. In place of the windmill system, a new smaller carbon foil holder will be mounted. There will be two silicon detectors (APIPS and PIPS detectors) mounted either side of a carbon foil holder, see Fig. 6.14. The holder will be attached to a linear actuator, allowing the height of the mount to be changed and the carbon foil in which to implant chosen. In addition to two carbon foils, there will be a copper plate, electrically connected to a BNC feedthrough at the side of the chamber, to be used as a Faraday cup and beam monitoring device. The detector assembly will be housed in a small aluminium rectangular chamber of 10 cm in diameter. The reduced dimensions from implantation site to germanium detector crystal (5 cm) will increase the solid angle of gamma-ray detection efficiency by at least a factor of 4, in addition to the aluminium walls providing less gamma-ray attenuation at low energy.

In addition, three germanium detectors will be easily mountable around the aluminium chamber. The small dimensions of the flange attaching the DSS to the

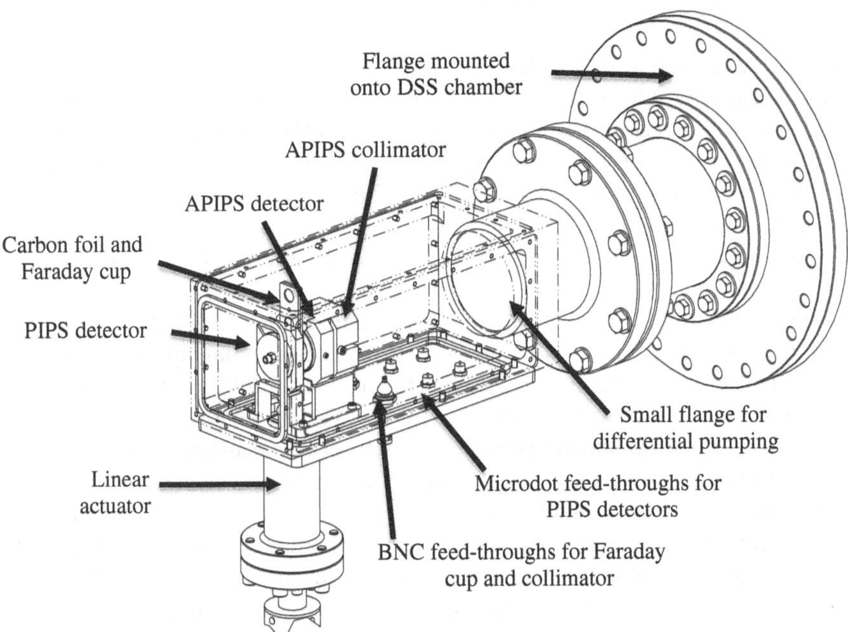

Fig. 6.13 Isometric drawing of the new design for the DSS

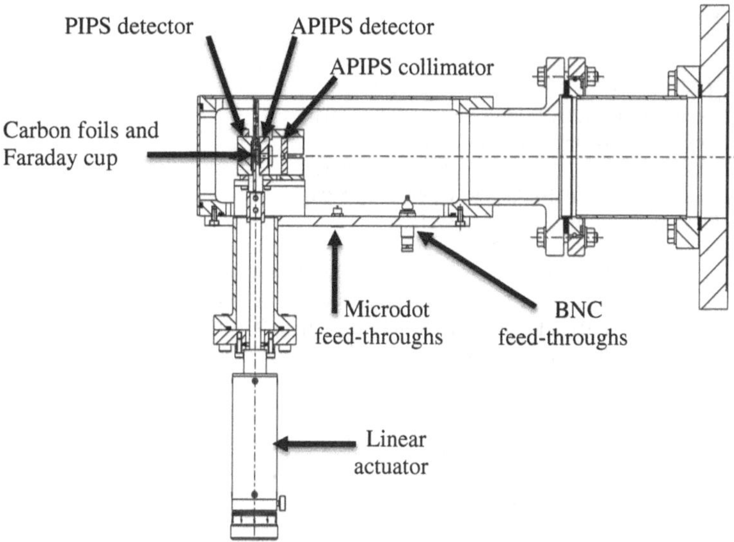

Fig. 6.14 Side view of the new design for the DSS

aluminium chamber will mean that, with the addition of a turbo pump in this location, the region will act as a differential pumping region, resulting in the (possibly higher, predicted to be around 10^{-7} mbar) vacuum in the aluminium chamber not affecting the better vacuum (10^{-9} mbar) in the rest of the CRIS beam line.

During the CRIS experiment, the digital data acquisition system for the DSS was set up so that only the energy of alpha particles could be measured by the PIPS detectors. This was a consequence of the high threshold applied to the signals from the silicon detector resulting from a fluctuating baseline (associated with the power of the 1,064 nm laser light). While conversion-electron detection is possible with the silicon PIPS detectors (by lowering the threshold parameters of the digital DAQ), the current setup would benefit from the installation of CZT detectors if the detection of internal transitions was required. Cadmium zinc telluride (CZT) detectors are direct band-gap semiconductors that can operate at room-temperature. They have a high sensitivity for x-rays and gamma rays (resulting from the high Z of cadmium and tellurium) and a better energy resolution than scintillation detectors [10]. Installation of such a detector in the DSS would allow the conversion electrons to be indirectly detected from the resulting characteristic x-rays. In addition, CZT crystals have a transparency in the mid-infra-red region of the electromagnetic spectrum, and it has been suggested that infra-red illumination at a wavelength of 1,020 nm improves the energy resolution at low temperatures [11]. This would indicate that the CZT detectors would be insensitive to the power of the 1,064 nm laser light that the PIPS detectors were so responsive to, and cooling the detectors (to between $-30\,°$C and $-15\,°$C) would further improve the energy resolution if needed.

References

1. Hurst GS et al (1979) Rev Mod Phys 51:767
2. Fedosseev VN et al (2012) Phys Scripta 85:058104
3. Letokhov V (1973) Opt Commun 7:59
4. P. Dendooven (1992) Ph.D. Thesis, IKS, KU Leuven
5. Andreyev AN et al (2010) Phys Rev Lett 105:252502
6. Cocolios TE et al (2010) J Phys G 37:125103
7. De Witte H et al (2007) Phys Rev Lett 98:112502
8. Lommel B et al (2002) Nucl Instrum Methods Phys Res A 480:199
9. Hennig W et al (2007) Nucl Instrum Methods Phys Res B 261:1000
10. CZT Technology (2011) Fundamentals and applications, GE Healthcare
11. Dorogov P et al (2012) IEEE Trans Nucl Sci 59:2375

Chapter 7
Spectroscopic Studies of Neutron-Deficient Francium

7.1 Collinear Resonance Ionization Spectroscopy

The neutron-deficient francium isotopes $^{202-207,211,220}$Fr were measured with collinear resonance ionization spectroscopy with respect to ^{221}Fr. The reasons for choosing the reference isotope ^{221}Fr were threefold. Firstly, the hyperfine structure of ^{221}Fr has been previously measured [1, 2] and the literature value is well established, thus providing an accurate consistency test for the results obtained. Secondly, the production mechanism of the ^{221}Fr isotope allowed reference hyperfine structure scans to be performed when protons were not impinging upon the UC_x target. This was due to the production of ^{225}Ac in the target-ion source (with a half-life of 10.0 days) decaying via alpha-particle emission to ^{221}Fr. This gave a constant ion beam of 30 pA of ^{221}Fr when the protons were off. This also avoided the concern of any intensity fluctuations associated with the fast release of francium from the target from direct proton production. The final consideration was the lifetime of the radioactive isotope. A half-life of 4.9 min meant limited radioactivity accumulated on the copper plate and hyperfine structure scans could be taken without the need to wait for the activity to decrease. The daughter activity of ^{213}Bi with a half-life of 45.6 min did however contribute significantly to the overall background, see Fig. 7.1. Despite this, the reference isotope of ^{221}Fr was favoured over ^{207,208}Fr because these isotopes have daughter isotopes with half-lives of 31.6 years (^{207}Bi) and 2.9 years (^{208}Po). Repeated reference measurements on either isotope would introduce long-lived contamination in the CRIS beam line. The number of the UC_x targets used during the August and October experiments were #477 and #490 respectively.

7.1.1 Consistency Analysis with ^{221}Fr

An example hyperfine structure scan of ^{221}Fr can be seen in Fig. 7.2. The number of resonant ions detected by the MCP is given as a function of laser frequency. As the

K.M. Lynch, *Laser Assisted Nuclear Decay Spectroscopy*,
Springer Theses, DOI: 10.1007/978-3-319-07112-1_7

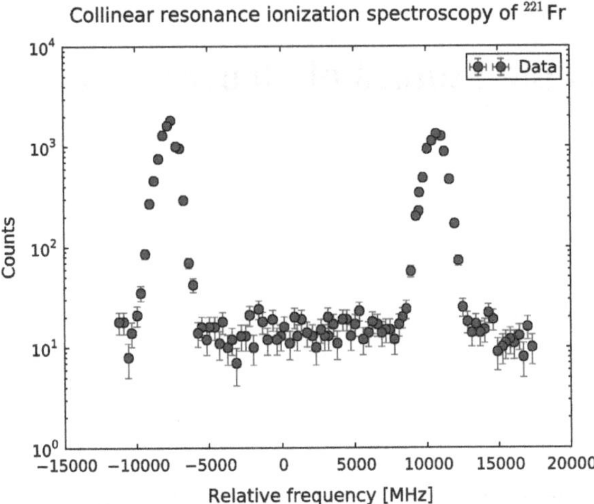

Fig. 7.1 The background of the hyperfine structure scan is attributed to the radioactive decay of ^{221}Fr and its daughter products

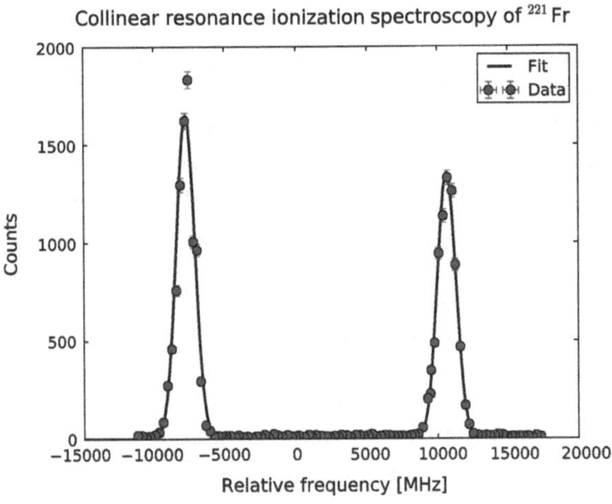

Fig. 7.2 An example hyperfine structure scan of the reference isotope ^{221}Fr

frequency of the resonant excitation step (the 422.7 nm light) is scanned, the hyperfine structure of ^{221}Fr is probed. A χ^2-minimization routine is used to fit the hyperfine structure spectrum, and the centroid frequency and hyperfine factors $A_{u,l}$ and $B_{u,l}$ are evaluated. As a reminder, the frequency of a transition between an upper and lower J level (J_{upper} and J_{lower} respectively) at which the hyperfine structure peak occurs is given by Eq. 2.8. The line profile of the hyperfine spectrum was determined to be purely Gaussian.

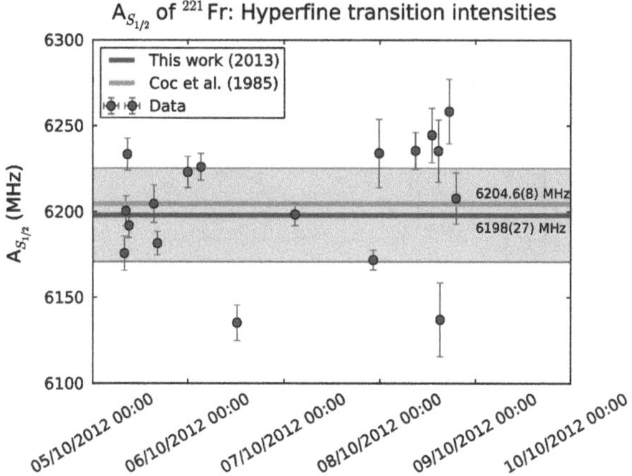

Fig. 7.3 Spread of $A_{S_{1/2}}$ values for ^{221}Fr when the intensities of the hyperfine peaks are calculated from angular momentum coupling considerations, given in Eq. 7.1. The weighted mean of 6,198(27) MHz agrees well with literature

Figure 7.1 shows the same scan with a logarithmic scale, displaying the background due to the radioactivity implanted on the copper plate. ^{221}Fr decays via alpha-particle emission to ^{217}At and ^{213}Bi. This implanted radioactivity β^- decays, emitting electrons (similar to when an ion impinged on the copper plate) which were detected by the MCP. This contributed significantly to the background of the ^{221}Fr scans due to the high yield. For francium isotopes with a lower rate of production and with shorter-lived decay products, this background was not observed once the ^{213}Bi activity had decayed away.

For the hyperfine structure of the neutron-deficient francium isotopes, the hyperfine factor $B_{P_{3/2}}$ is small enough to be considered negligible. In addition, when $\mathrm{J} = 1/2$ (the case for the lower state 7s $^2\mathrm{S}_{1/2}$), the hyperfine factor $B_{S_{1/2}} = 0$, as defined in Eq. 2.5. For this reason, the B factors were set to zero in the χ^2-minimization fitting. In addition, the $A_{P_{3/2}}$ value was fixed to the ratio of the 7s $^2\mathrm{S}_{1/2} \rightarrow$ 8p $^2\mathrm{P}_{3/2}$ transition given in literature [2]

$$\frac{A_{P_{3/2}}}{A_{S_{1/2}}} = \frac{B_{e,(0),P_{3/2}}}{J_{P_{3/2}}} \frac{J_{S_{1/2}}}{B_{e,(0),S_{1/2}}} = \frac{22.4}{6{,}209.9}.$$

The calculated hyperfine factor $A_{S_{1/2}}$ for all reference scans of ^{221}Fr are shown in Fig. 7.3. The weighted mean for $A_{S_{1/2}}$ was calculated to be 6,198(27) MHz and is shown by the blue line, with the grey region representing a confidence interval of 68 % (one standard deviation). This value agrees within errors of the established literature value of 6,204.6(8) MHz [1].

The intensities of the hyperfine transitions $S_{F_{upper} F_{lower}}$ between hyperfine levels F_{lower} and F_{upper} (with angular momentum J_{lower} and J_{upper} respectively) are related to the intensity of the underlying fine structure transition $S_{J_{upper} J_{lower}}$ [3].

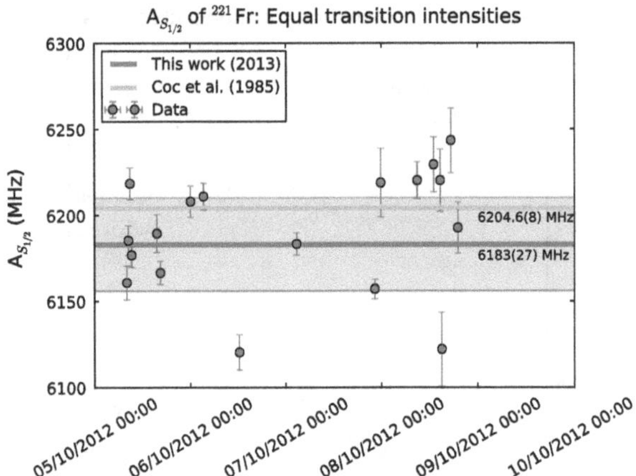

Fig. 7.4 Spread of $A_{S_{1/2}}$ values for ^{221}Fr when the hyperfine peaks are set to equal intensities. Although within errors, the weighted mean deviates from literature by 20 MHz

Therefore the relative intensities of the hyperfine transitions can be given by

$$\frac{S_{F_{upper}F_{lower}}}{S_{J_{upper}J_{lower}}} = (2F_{lower}+1)(2F_{upper}+1)\begin{Bmatrix} F_{lower} & F_{upper} & 1 \\ J_{upper} & J_{lower} & I \end{Bmatrix}^2, \tag{7.1}$$

where $\{\ldots\}$ denotes the Wigner 6-j coefficient. Figure 2.2 in Sect. 2.1 shows a simulated hyperfine structure for ^{221}Fr when this equation is implemented to calculate the intensities of the hyperfine structure peaks. The lower state splitting of the 7s $^2S_{1/2}$ state is ~18 GHz. The upper state splitting of the 8p $^2P_{3/2}$ state is ~300 MHz. This splitting could not be resolved as the line width of the laser was 1.5 GHz.

Figure 7.3 shows the $A_{S_{1/2}}$ values when Eq. 7.1 is implemented to calculate the intensities of the hyperfine structure peaks. In contrast, Fig. 7.4 shows the values of $A_{S_{1/2}}$ obtained when angular momentum couplings are not taken into consideration, and the intensities of the peaks are equalized. Due to the modification this produces in the centre of gravity of each peak envelope, a smaller $A_{S_{1/2}}$ value of 6,183(27) MHz is calculated. This illustrates that while the current experiment could not resolve the upper state splitting, its presence causes a systematic shift in the centroid position compared to $A_{P_{3/2}}$.

The centroid frequency of the reference scans are shown in Fig. 7.5. The weighted mean was calculated to be −228(79) MHz. There is a large scatter from one hyperfine structure scan to the next, in addition to a long-term drift over the course of the experiment. The scatter of the centroid frequency is attributed to the variation in synchronization of the two laser pulses in the resonance ionization process. The drift may be due to the changes in atmospheric conditions, movement of the optical fibre coupling or the calibration of the wavemeter. To account for this, a linear fit to the data in time was calculated, shown in Fig. 7.6. Instead of calculating the isotope shifts using the global value of −228(79) MHz for ^{221}Fr, a local value is determined.

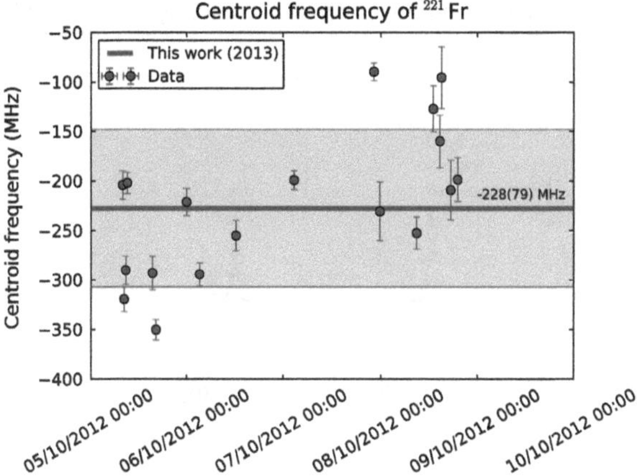

Fig. 7.5 Spread of centroid frequency values for ^{221}Fr. The large scatter is attributed to the variation in synchronization between the two laser pulses in the RIS process

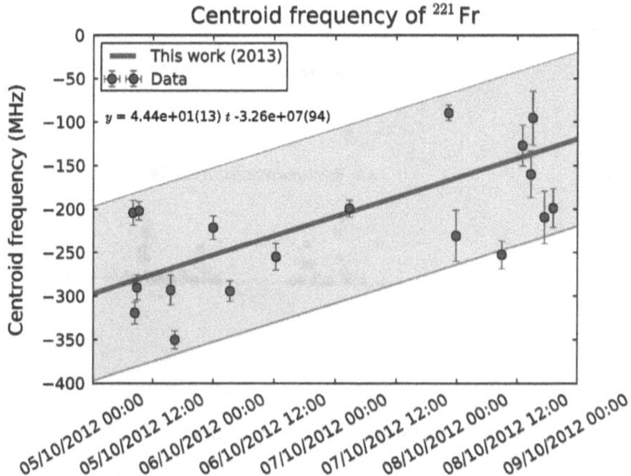

Fig. 7.6 A linear fit to the centroid frequency values for ^{221}Fr accounts for the drift in centroid frequency with time. An uncertainty of 50 MHz is attributed to the linear fit of the data

The value for the centroid frequency is calculated from the time each scan of an isotope was taken, using the linear fit

$$f(x) = P[0]t + P[1], \tag{7.2}$$

where t is the time of the scan, to account for the drift in the frequency. This allows for a more accurate determination of isotope shifts for isotopes measured after the last ^{221}Fr scan. The systematic error from this approach was evaluated to be 50 MHz.

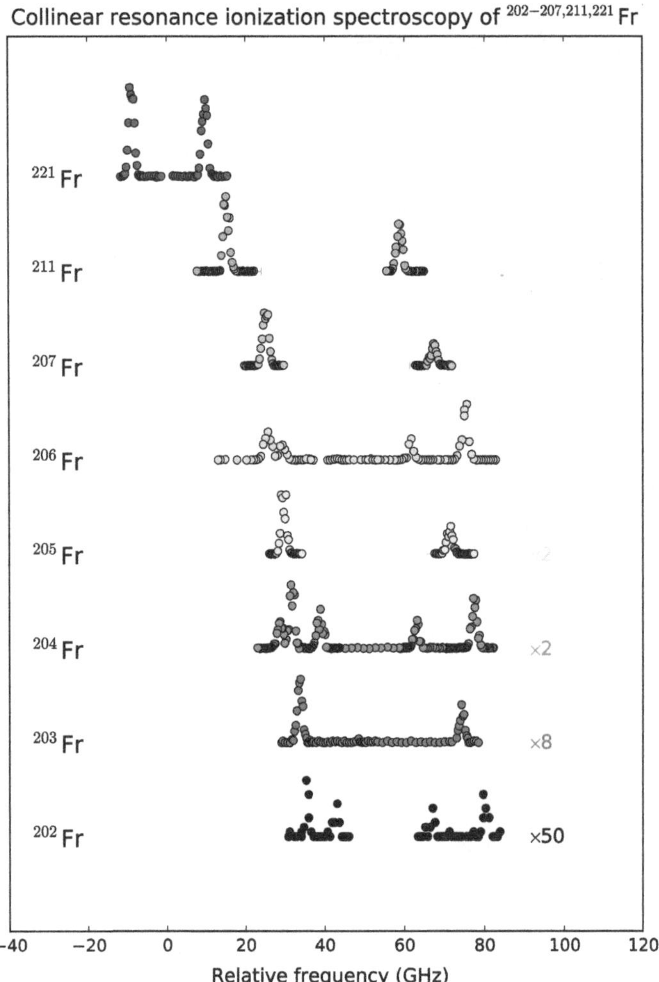

Fig. 7.7 Collinear resonance ionization spectroscopy on $^{202-207,211,221}$Fr

7.1.2 Collinear Resonance Ionization Spectroscopy of $^{202-207}$Fr

The hyperfine structure of the neutron-deficient francium isotopes $^{202-207,211,220,221}$Fr were measured with collinear resonance ionization spectroscopy. A plot of the hyperfine structure scans of these isotopes can be seen in Fig. 7.7, illustrating the isotope shift of the francium isotopes as they become more neutron-deficient. Example scans of the even-A francium isotopes ^{202}Fr, ^{204}Fr and ^{206}Fr are shown in the following chapter (Figs. 7.8, 7.16 and 7.26 respectively). The solid lines are colour-coded to indicate the fit used to extract the hyperfine factors for individual states: (3^+), (7^+) and (10^-). The summed fit to the data for the even-A francium

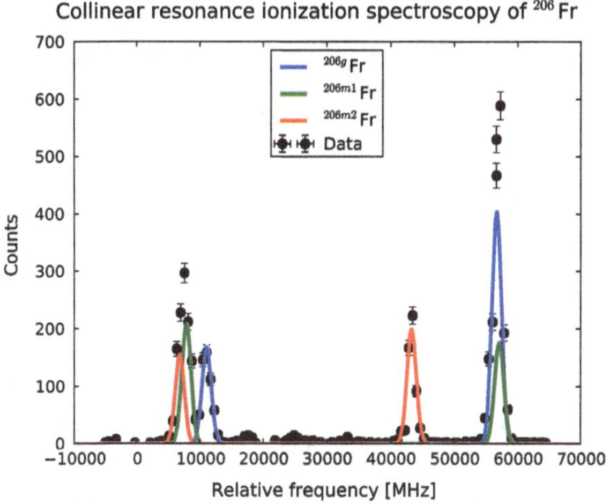

Fig. 7.8 Collinear resonance ionization spectroscopy scan of ^{206}Fr

isotopes are presented in Appendix A, alongside example scans of the odd-A francium isotopes.

The $A_{S_{1/2}}$ factor and the centroid frequency of the hyperfine structure was determined for each scan individually. For isotopes with multiple scans, a weighted mean for the $A_{S_{1/2}}$ factor and the centroid frequency was calculated based on the error of the fits. The error was calculated as the weighted standard deviation of the values. The isotope shifts were determined relative to ^{221}Fr, where the value of the centroid frequency of ^{221}Fr was calculated from the time the scan was taken, see Sect. 7.1.1.

7.1.3 Determination of the ^{206}Fr Hyperfine Spectra

Two sets of data were used in the determination of the nuclear observables from the hyperfine structure scans. The data set for the francium isotopes $^{202-207}$Fr was taken in August 2012, while the data set for $^{202-205,207,211,218-220}$Fr was taken in October 2012. While the value for the $A_{S_{1/2}}$ factor is an independent measurement, the isotope shift is intrinsically related to the relative frequency of ^{221}Fr. The hyperfine structure scan of ^{206}Fr, taken in August 2012, is shown in Fig. 7.8. This was the only measurement of the isotope obtained during the experimental campaign, so the data for ^{206}Fr needed to be calibrated with that of October to correctly analyse the hyperfine structure.

Figure 7.9 shows the isotope shifts of francium taken in August and October 2012. In order to tie the value of ^{206}Fr from August to the October run, the data from each experiment was plotted against each other, shown in Fig. 7.10. A linear fit was made to the data,

$$y = 1.003(2)x, \tag{7.3}$$

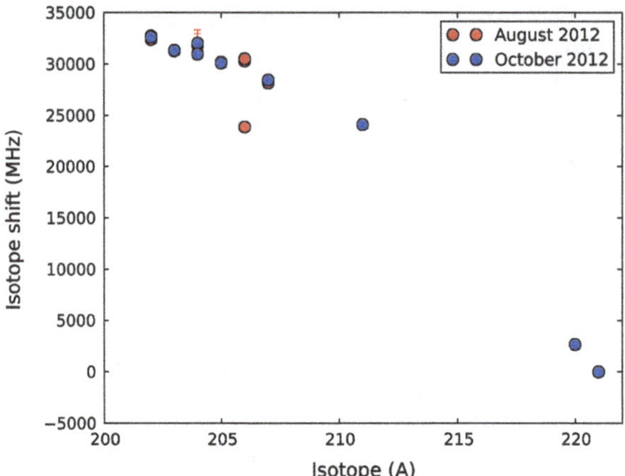

Fig. 7.9 The August and October data displays a shift in frequency of the isotope shifts

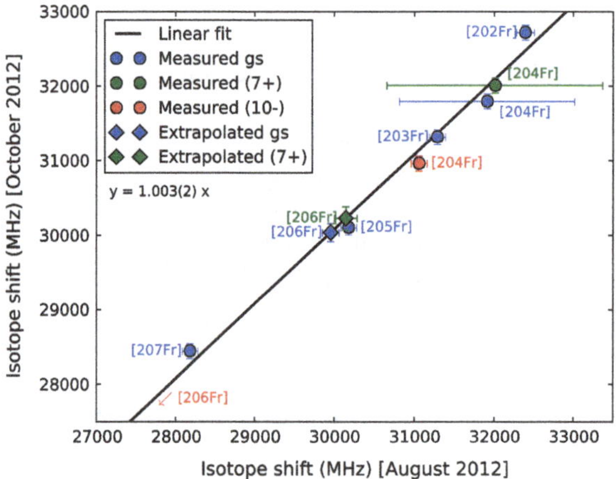

Fig. 7.10 The isotope shifts from August and October data were plotted. A linear equation was fit to the data, which allowed extrapolation of the August ^{206}Fr data to the October frequency

which allowed the isotope shifts of ^{206}Fr (from August 2012) to be calibrated for inclusion with the October data set. The values are shown in Fig. 7.10, where the intercept was set to zero, corresponding to the isotope shift of ^{221}Fr for both August and October data. This allowed the isotope, and isomer, shifts of ^{206g,m1,m2}Fr to be evaluated with respect to the rest of the data from October.

The measurement of the hyperfine spectrum of ^{206}Fr led to the question of identification of peaks. Recent measurements of the neutron-deficient francium isotopes at TRIUMF [4] identified the ground and first isomeric state of ^{206}Fr (spin 3^+ and 7^+

respectively), but did not observe ^{206m2}Fr (spin 10^-). This may be attributed to the lower rate of production and shorter half-life (0.7 s compared to 16 s for the positive parity states). From this information, it is possible to assign the first right-hand peak (labeled D) to ^{206m2}Fr and the second left-hand peak (labeled C) to either ^{206g}Fr or ^{206m1}Fr. These two options can be seen in Figs. 7.11a and 7.12a: peak C is identified as ^{206g}Fr in Fig. 7.11a and ^{206m1}Fr in Fig. 7.12a.

Given these two possibilities, the hyperfine structure was evaluated for all possible options, which are presented in Tables 7.1 and 7.2. When peak C is attributed to ^{206g}Fr, one possible fit (Option 1a) is shown in Fig. 7.11a. The unresolved peaks of A and B, and E and F, means that each peak can either be on the left- or right-hand side. This results in four different fits to option 1, given in Table 7.1. These fits were evaluated and their charge radii presented in Fig. 7.11(Below). The same method was repeated for option 2, where peak C is attributed to ^{202m1}Fr. One possible fit (Option 2a) is shown in Fig. 7.12a. The four fitting options (presented in Table 7.2) were evaluated and their charge radii presented in Fig. 7.12(Below).

From these results, identification of the peaks based on the systematics of the region can occur. In addition to the information about the identity of the ^{206}Fr peaks, the TRIUMF data also provided the knowledge that the charge radii of ^{206g}Fr, in comparison to the other francium isotopes, had the same characteristics as the lead isotopes in that region. This allowed option 2 to be discounted, due to the reversal in odd-even staggering resulting from the configuration. Based on the systematics of the ^{204}Fr states, option 1b became the most likely arrangement of the hyperfine structure. In ^{204}Fr, the charge radii of the first isomeric state (green) is below the ground state (blue). For ^{206}Fr, the only option with a similar result is option 1b, thus this arrangement is preferred for these states.

The following analysis of ^{206}Fr and its relation to the other francium isotopes is based on this assumption. Further investigation of the hyperfine structure of ^{206}Fr is clearly needed to prove the existence of the second isomeric state, which was absent in the TRIUMF data, see Fig. 7.8. The separation of approximately 13–14 GHz between the hyperfine peaks of ^{206m2}Fr and ^{206g,m1}Fr on the right-hand side of the spectrum could be a result of the laser multi-moding, and would explain both experiments' results. The difference in frequency of the two peaks equals the mode spacing of the Ti:Sa laser (13.7 GHz), producing the light for the resonant ionization excitation step [5].

However, the difference in frequency of the two peaks on the right-hand side of the spectrum of ^{204}Fr in Fig. 7.16 is similarly between 13.5 and 14.5 GHz. Since the presence of the three isomers in ^{204}Fr has been confirmed (see Sect. 7.2.1), the peak in Fig. 7.8 is consequently tentatively identified as ^{206m2}Fr. It should be noted that only one scan was taken for ^{206}Fr, but in all other scans during the experiment in August, no effect of such magnitude was observed. The data were thus analysed and interpreted based on the tentative identification of the ^{206m2}Fr state in the spectrum, with the caveat that more hyperfine structure studies of ^{206}Fr are needed to fully characterise ^{206}Fr and its isomer(s).

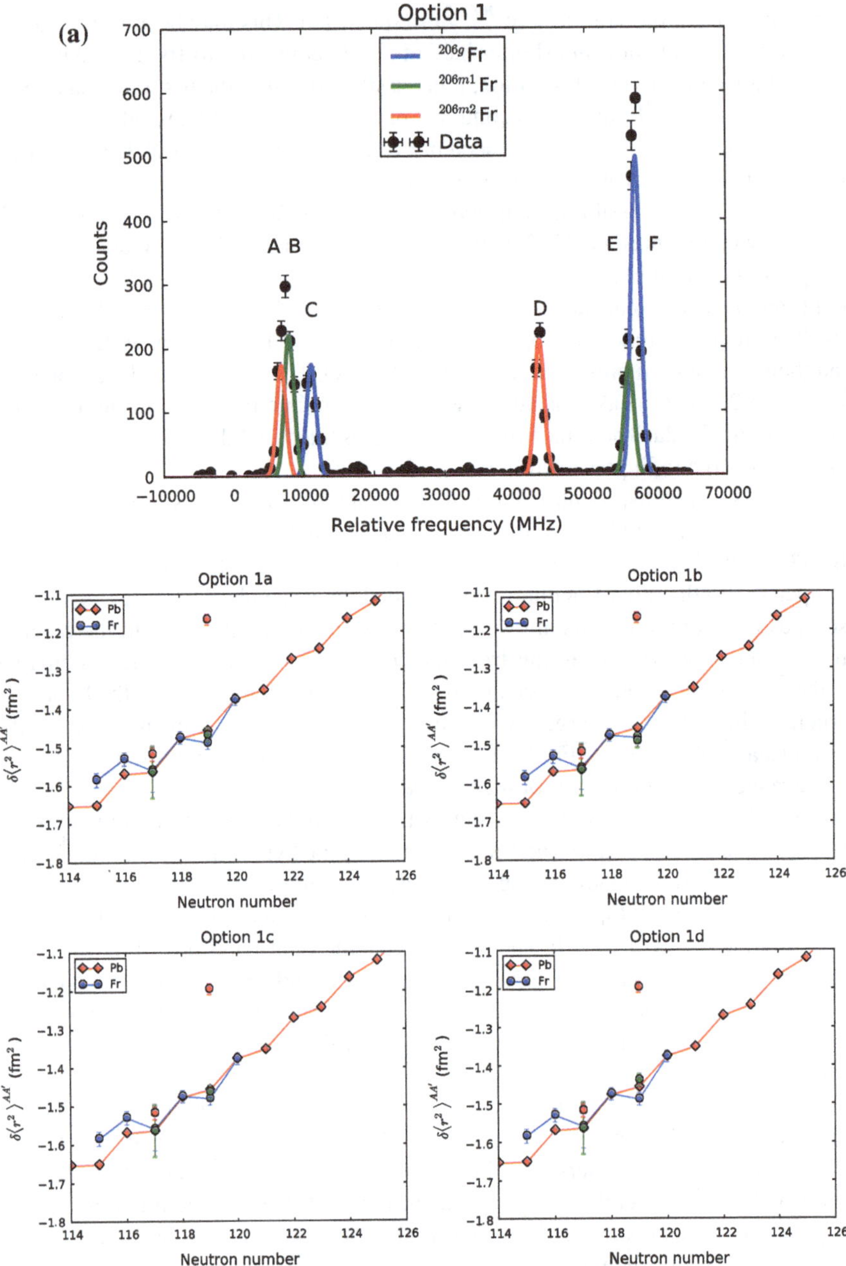

Fig. 7.11 The charge radii resulting from the different hyperfine structure possibilities of fitting option 1, given in Table 7.1. Peak C is assigned to ^{206g}Fr

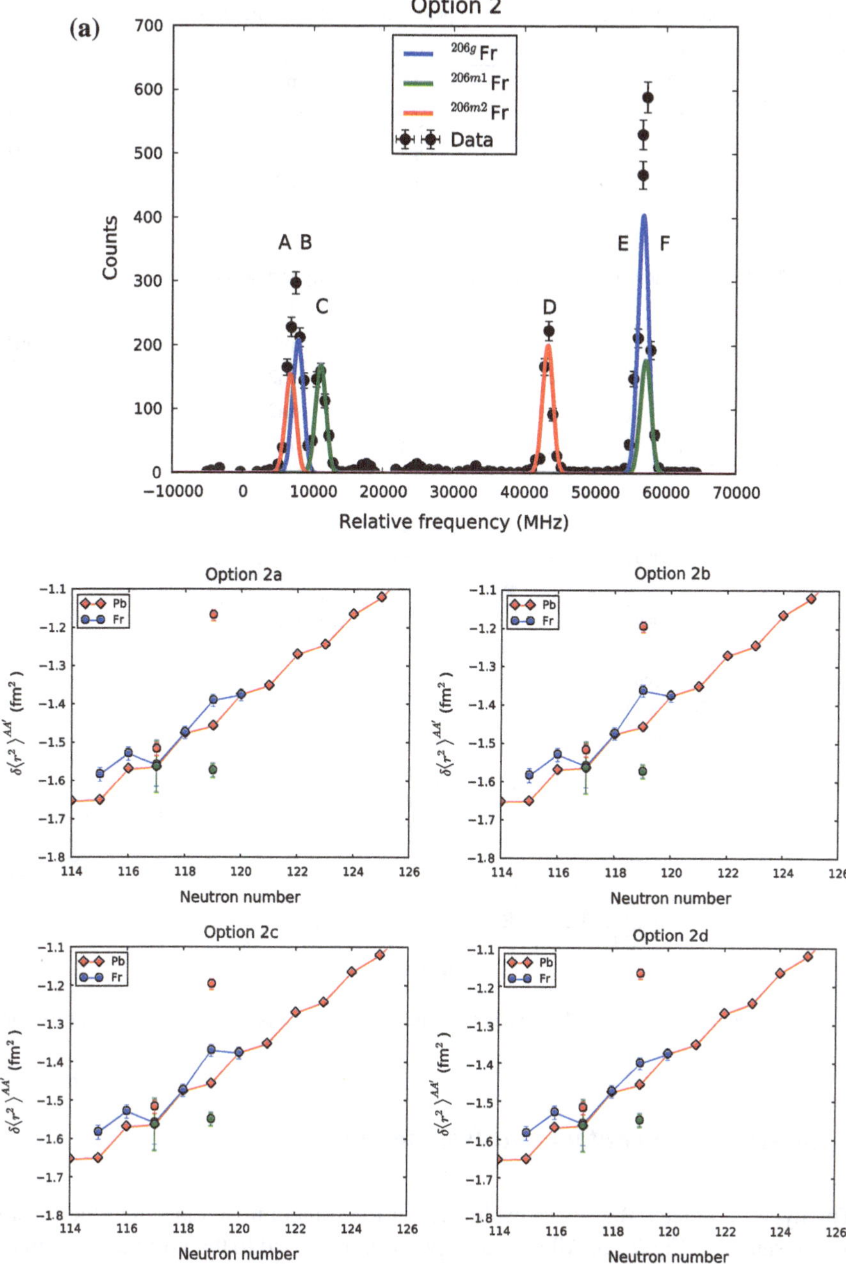

Fig. 7.12 The charge radii resulting from the different hyperfine structure possibilities of fitting option 2, given in Table 7.2. Peak C is assigned to ^{206m1}Fr

Table 7.1 Option 1: Identification options of peaks in ^{206}Fr hyperfine structure

Option	A	B	C	D	E	F
1a	m2	m1	g	m2	m1	g
1b	m2	m1	g	m2	g	m1
1c	m1	m2	g	m2	g	m1
1d	m1	m2	g	m2	m1	g

Table 7.2 Option 2: Identification options of peaks in ^{206}Fr hyperfine structure

Option	A	B	C	D	E	F
2a	m2	g	m1	m2	g	m1
2b	g	m2	m1	m2	g	m1
2c	g	m2	m1	m2	m1	g
2d	m2	g	m1	m2	m1	g

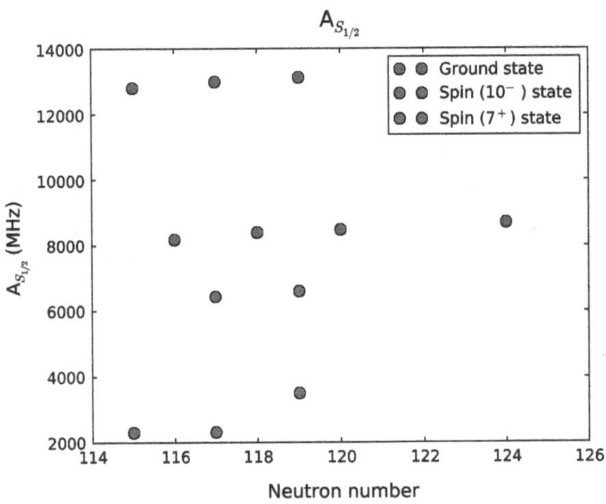

Fig. 7.13 The hyperfine factor $A_{S_{1/2}}$ values for the francium isotopes $^{202-207,211}$Fr. Error bars are within the data points

7.1.4 Hyperfine Factors and Isotope Shifts

The hyperfine factor $A_{S_{1/2}}$ was evaluated for the francium isotopes, and the results are presented in Fig. 7.13. All blue data points correspond to the ground state, green data points show the spin 7^+ isomeric states (present in ^{204,206}Fr) and red data points show the spin 10^- isomeric states (present in ^{202,204,206}Fr).

Table 7.3 presents a comparison of the $A_{S_{1/2}}$ factor values measured by CRIS and Coc et al. [1]. From the standard deviation of the $A_{S_{1/2}}$ factor values for the reference isotope ^{221}Fr (see Fig. 7.3) of 27 MHz, a conservative error on all $A_{S_{1/2}}$ factor values

Table 7.3 Comparison of measured $A_{S_{1/2}}$ values with literature

Isotope	Spin	$A_{S_{1/2}}$ (MHz) [CRIS]	$A_{S_{1/2}}$ (MHz) [1]
207	$9/2^-$	+8480(30)	+8484(1)
211	$9/2^-$	+8696(63)	+8713.9(8)
220	1^-	−6505(42)	−6549.4(9)
221	$9/2^-$	+6198(30)	+6204.6(8)

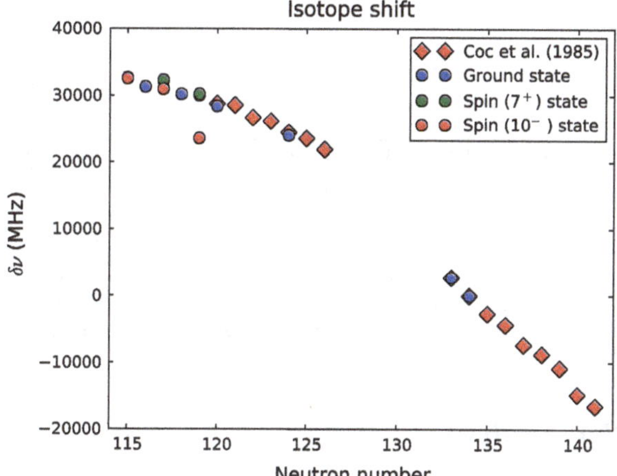

Fig. 7.14 The isotope shifts for the francium isotopes $^{202-207,211,218-221}$Fr, relative to ^{221}Fr. Results from this work are combined with isotope shifts measured by Coc et al. [1]. Error bars are within the data points

of at least 30 MHz was determined. The significantly smaller errors attributed to the Coc values are due to their ability to control the frequency of the laser (for the resonant step) with an accuracy of 1 MHz and a frequency resolution of the lasers of less than 1 MHz. This is performed step-by-step by a sigmameter which provided a linear frequency scale. Thus the upper state splitting is resolved and the error on the $A_{S_{1/2}}$ factor is smaller.

The isotope shifts for the neutron-deficient francium isotopes measured to date are shown in Fig. 7.14. In addition to the results obtained from this work (circle data points), the previously measured isotope shifts for $^{207-213,220-228}$Fr by Coc et al. [1] are presented (diamond data points). Originally with reference to ^{212}Fr, these results were re-scaled relative to ^{221}Fr. Coc probes the 7s $^2S_{1/2} \rightarrow$ 7p $^2P_{3/2}$ transition, whereas this work excited the electron from the 7s $^2S_{1/2}$ state to the 8p $^2P_{3/2}$ state. Despite probing a different transition, the isotope shifts are remarkably similar, see Fig. 7.14. This highlights the dominance of the 7s $^2S_{1/2}$ state in probing the nuclear volume over the 7p $^2P_{3/2}$ or 8p $^2P_{3/2}$ states. Figure 7.14 also shows the global change in isotope shifts as the N = 126 shell closure is crossed, resulting from the change in the nuclear volume.

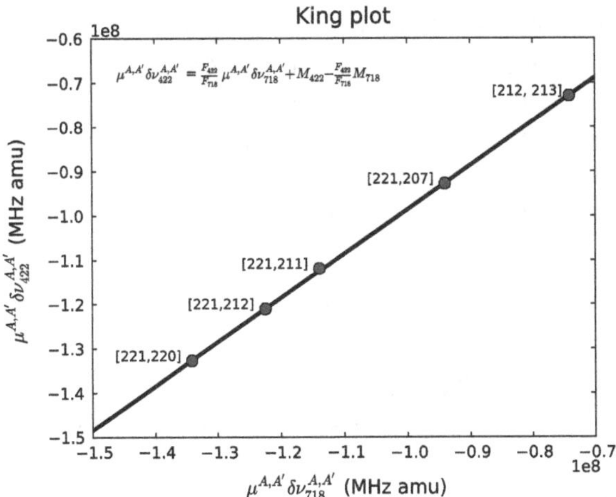

Fig. 7.15 King plot of 718nm and 422.7 nm transitions for ^{207,211,212,220}Fr with reference to ^{221}Fr, and the ^{213}Fr transition with reference to ^{212}Fr. The values for the 718 nm transition were taken from Coc et al. [1]. The values for the 422.7 nm transition are from CRIS (^{207,211}Fr) and Duong (^{212,213,220}Fr) [2]. The reduced $\chi^2 = 0.6$. Error bars are within the data points

7.1.5 King Plot Analysis

In order to extract the charge radii from the isotope shifts, the King plot method was used, as outline in Sect. 2.2.4. This combines the previously measured isotope shifts by Coc et al. [1] probing the 7s $^2S_{1/2} \rightarrow$ 7p $^2P_{3/2}$ transition with 718 nm laser light, with those made by Duong et al. [2] probing the 7s $^2S_{1/2} \rightarrow$ 8p $^2P_{3/2}$ transition with 422.7 nm light. The isotope shifts of ^{207,211}Fr (with respect to ^{221}Fr) from this experiment were combined with the isotope shifts of ^{220,213}Fr (with respect to ^{221}Fr and ^{212}Fr) from Duong. These values were plotted against the corresponding isotope shifts from Coc et al. [1]. This represents the first time the 7s $^2S_{1/2} \rightarrow$ 8p $^2P_{3/2}$ transition using 422.7 nm laser light has been calibrated. From the gradient and intercept of the resulting fit, the atomic factors can be extracted from Eq. 2.26. These were found to be

$$\text{Gradient} = \frac{F_{422}}{F_{718}} = 0.995(3),$$

$$\text{Intercept} = M_{422} - \frac{F_{422}}{F_{718}} M_{718} = 837(308)\ \text{GHz amu}.$$

From these values, the atomic factors for the 422 nm transition was calculated to be

$$F_{422} = -20.668(214)\ \text{GHz/fm}^2,$$

Table 7.4 Experimental isotope shift, mass shift and field shift measurements between ^{221}Fr and the neutron-deficient francium isotopes

Isotope	Spin	Expt. IS (MHz)	MS (MHz)	FS (MHz)
202g	(3^+)	32680(100)	−320(140)	33000(170)
202m	(10^-)	32570(130)	−320(140)	32890(190)
203	$(9/2^-)$	31300(100)	−300(130)	31620(170)
204g	(3^+)	32190(100)	−280(120)	32480(160)
204m1	(7^+)	32320(100)	−280(120)	32600(160)
204m2	(10^-)	30990(100)	−280(120)	31270(160)
205	$(9/2^-)$	30210(100)	−270(120)	30480(150)
206g	(3^+)	30040(120)	−250(110)	30290(160)
206m1	(7^+)	30230(160)	−250(110)	30480(190)
206m2	(10^-)	23570(120)	−250(110)	23820(160)
207	$9/2^-$	28420(100)	−230(100)	28660(140)
211	$9/2^-$	24040(100)	−160(70)	24200(120)
220	$5/2^-$	2750(100)	−15(7)	2770(100)

$$M_{422} = +752(327)\ \text{GHz amu}.$$

For comparison, the atomic factors evaluated for the 718 nm transition were determined by Dzuba to be $F_{718} = -20.766(208)\,\text{GHz/fm}^2$ and $M_{718} = -85(113)\,\text{GHz}$ amu [6]. The mass factor (given in Eq. 2.19) is the combination of two components: the normal mass shift K_{NMS} and the specific mass shift K_{SMS}; and is dependent on the frequency of the transition probed. The value of the normal mass shift can be calculated for the 422.7 nm transition from the equation given by Dzuba et al. [6], and then calculation of the specific mass shift becomes possible,

$$K_{NMS} = \frac{\nu_{exp}}{1822.888} = +389\ \text{GHz amu},$$

$$K_{SMS} = M_{422} - K_{NMS} = +363(327)\ \text{GHz amu}.$$

These mass shift factors can be used to calculated the normal mass shift and specific mass shift contributions to the mass shift.

The total mass shift $\delta\nu_{MS}$ and field shift $\delta\nu_{FS}$ can be determined by

$$\delta\nu_{MS} = M_{422}\frac{A' - A}{AA'}, \tag{7.4}$$

and,

$$\delta\nu_{FS} = \delta\nu_{IS} - \delta\nu_{MS}, \tag{7.5}$$

Table 7.5 Comparison of extracted charge radii with literature

Isotope	Spin	$\delta\langle r^2\rangle$ (fm^2) [CRIS]	$\delta\langle r^2\rangle$ (fm^2) [6]
207	9/2$^-$	−1.386(16)	−1.386(11)
211	9/2$^-$	−1.171(13)	−1.178(12)
220	1$^-$	−0.134(5)	−0.133(14)

where A is the mass of the reference isotope ^{221}Fr and A' is the mass of the francium isotope under investigation. The mass shift and field shift results are given in Table 7.4.

7.1.6 Changes in Mean-Square Charge Radii

By rearrangement of Eq. 2.23, the change in mean-square charge radii between isotopes can be determined. Table 7.5 presents the charge radii extracted for the francium isotopes ^{207,211,220}Fr from this work compared to those found in literature [6]. The experimental error on the CRIS values results from the uncertainty on the experimental isotope shift and the analysis of the atomic factors. The increasing magnitude of the errors with decreasing mass arises from the uncertainty in the calculation of the atomic factors for the 7s ^{2}S$_{1/2}$ → 8p ^{2}P$_{3/2}$ transition. The mass factor M_{422} specifically has a large associated uncertainty, dominating the propagation of error for the charge radii. The charge radii values re-evaluated by Dzuba were deduced from the 7s ^{2}S$_{1/2}$ → 7p ^{2}P$_{3/2}$ isotope shifts. Due to the high accuracy of these experimental isotope shifts, the errors are dominated by the theoretical calculation of the field factor F_{718} of the atomic factors.

Excellent agreement can be seen between the CRIS results and literature, with a deviation of only 0.5 % for ^{211}Fr. The smaller error now attributed to the charge radii for ^{220}Fr is due to the smaller extrapolated uncertainty on the mass factor, so close to the reference isotope ^{221}Fr.

7.1.7 Magnetic Dipole Moments

The magnetic moment is extracted from the hyperfine factor $A_{S_{1/2}}$ using the known magnetic moment of a reference isotope, as defined in Eq. 2.11. For the francium isotopes, the magnetic moments were calculated using two reference measurements: ^{211}Fr and ^{210}Fr, as shown in Table 7.6.

The third column of Table 7.6 presents the magnetic moments evaluated from the CRIS data with reference to the magnetic moment of ^{211}Fr. This was measured to be $\mu(^{211}\text{Fr}) = 4.00(8)\,\mu_N$ by Ekström et al. in 1986 [7]. The fourth column presents the magnetic moments of Ekström available from literature [7]. The experimental error on the CRIS-Ekström values are solely from the error associated with the hyperfine

Table 7.6 Comparison of extracted magnetic dipole moments with literature

Isotope	Spin	μ (μ_N) [CRIS][a]	μ (μ_N) [7][a]	μ (μ_N) [CRIS][b]	μ (μ_N) [7][b]
207	$9/2^-$	+3.90(8)	+3.89(8)	+3.87(5)	+3.87(5)
211	$9/2^-$	+4.00(9)	+4.00(8)	+3.97(5)	+3.98(5)
220	1^-	−0.66(1)	−0.67(1)	−0.66(1)	−0.66(1)
221	$5/2^-$	+1.58(3)	+1.58(3)	+1.57(2)	+1.57(2)

The third and fourth columns present the magnetic moments of CRIS and Ekström evaluated with respect the magnetic moment of ^{211}Fr [7]. The fifth and sixth columns tabulate the same magnetic moments re-evaluated with respect to the more accurate magnetic moment of ^{210}Fr [8]

[a] Evaluated with the magnetic moment of ^{211}Fr of Ekström et al. [7]

[b] Evaluated with the magnetic moment of ^{210}Fr of Gomez et al. [8]

factor $A_{S_{1/2}}$ propagated with the error on the value of the magnetic moment of the reference isotope. Again, there is excellent agreement with the CRIS-Ekström values to those of literature.

The fifth column presents the CRIS magnetic moments evaluated with reference to the most recent (and accurate) evaluation of a francium isotope's magnetic moment to date. The magnetic moment of ^{210}Fr was determined to be $\mu(^{210}\text{Fr}) = 4.38(5)\,\mu_\text{N}$ by Gomez in 2008 [8]. This work measured the hyperfine splitting of the 9s $^2\text{S}_{1/2}$ level of ^{210}Fr which has lower correlation effects than the 7s $^2\text{S}_{1/2}$ state. Agreement between the CRIS-Gomez values and literature is good, with the smaller error a result of the more accurate measurement of the reference magnetic moment. The sixth column presents the magnetic moments of Ekström, re-evaluated with respect to the new reference magnetic moment of ^{210}Fr of Gomez. There is very good agreement between the CRIS-Gomez and the Ekström-Gomez values. This highlights the need for a re-evaluation of literature values with reference to this new, and more accurate, measurement of the magnetic moment of ^{210}Fr.

7.1.8 Nuclear Observables from Hyperfine Spectra

Table 7.7 shows the hyperfine factor, isotope shift, charge radii and magnetic moment values extracted from the CRIS data for the francium isotopes $^{202-207,211,220}$Fr with reference to ^{221}Fr. The hyperfine factor $A_{S_{1/2}}$ is the weighted mean of $A_{S_{1/2}}$ values for isotopes where more than one hyperfine structure scan is present. The error associated with this value is the weighted standard deviation of these values. Where only one scan is available, the error is taken as the uncertainty on $A_{S_{1/2}}$ (calculated from the fit of the data). A minimum error of 30 MHz was attributed to the $A_{S_{1/2}}$ factor values due to the scatter of the measured $A_{S_{1/2}}$ for the reference isotope ^{221}Fr.

The isotope shift values were calculated in a similar way, as the weighted mean of all isotope shifts for a given isotope. The error on the isotope shift was determined to be 100 MHz due to the long-term drift of the centroid frequency of ^{221}Fr as the experiment progressed, and the scan-to-scan scatter in centroid frequency. When the

Table 7.7 The nuclear observables extracted from the hyperfine structure scans of the neutron-deficient francium isotopes $^{202-207,211,220}$Fr with reference to ^{221}Fr

Isotope	Spin	$A_{S_{1/2}}$ (MHz)	Isotope shift (MHz)	Charge radii (fm^2)	Magnetic moment (μ_N)
202g	(3$^+$)	+12800(50)	32680(100)	−1.596(18)	+3.90(5)
202m	(10$^-$)	+2300(30)	32570(130)	−1.591(19)	+2.34(4)
203	(9/2$^-$)	+8180(30)	31320(100)	−1.530(18)	+3.73(4)
204g[a]	(3$^+$)	+12990(30)	32190(100)	−1.571(18)	+3.95(5)
204m1[a]	(7$^+$)	+6440(30)	32320(100)	−1.577(18)	+4.57(6)
204m2[a]	(10$^-$)	+2310(30)	30990(100)	−1.513(17)	+2.35(4)
205	(9/2$^-$)	+8400(30)	30210(100)	−1.475(17)	+3.83(5)
206g	(3$^+$)	+13120(30)	30040(120)	−1.465(17)	+3.99(5)
206m1	(7$^+$)	+6610(30)	30230(160)	−1.475(18)	+4.69(6)
206m2	(10$^-$)	+3500(30)	23570(120)	−1.153(14)	+3.55(4)
207	9/2$^-$	+8480(30)	28420(100)	−1.386(16)	+3.87(5)
211	9/2$^-$	+8700(60)	24040(100)	−1.171(13)	+3.97(5)
220	1$^-$	−6500(40)	2750(100)	−0.134(5)	−0.66(1)
221	5/2$^-$	+6200(30)	0(0)	0.000(0)	+1.57(2)

The values listed for ^{204}Fr have been calculated from the alpha-gated hyperfine structure scans, which is discussed in Sect. 7.2.2. This technique gives the ability to disentangle overlapping hyperfine structures to extract their nuclear observables with a much greater degree of accuracy
[a] Calculated from the alpha-decay gated hyperfine structure scan of ^{204}Fr

calculated weighted standard deviation of the isotope shift was higher than 100 MHz, this error in quoted instead.

The calculation of the magnetic moments was evaluated in reference to the magnetic moment of ^{210}Fr, measured by Gomez et al. [8]. As discussed in Sect. 7.1.7, this is the most accurate measurement to date, which also highlights the need for a re-evaluation of magnetic moments of the francium isotopes in literature.

The values for ^{204}Fr and its isomers were not calculated from CRIS ion data, but instead from alpha-decay gated hyperfine structure scans. More information on how this data was collected and analysed can be found in Sect. 7.2.2.

7.2 Laser Assisted Nuclear Decay Spectroscopy

7.2.1 Laser Assisted Nuclear Decay Spectroscopy of ^{204}Fr

The hyperfine structure of ^{204}Fr and its isomers are shown in Fig. 7.16. The identification of the three states was performed with laser assisted nuclear decay spectroscopy. The laser was tuned on resonance with each of the first three hyperfine resonances, and decay spectroscopy was performed on each. A 20 min decay measurement was performed on all three states, for alpha-particle and gamma-ray emission.

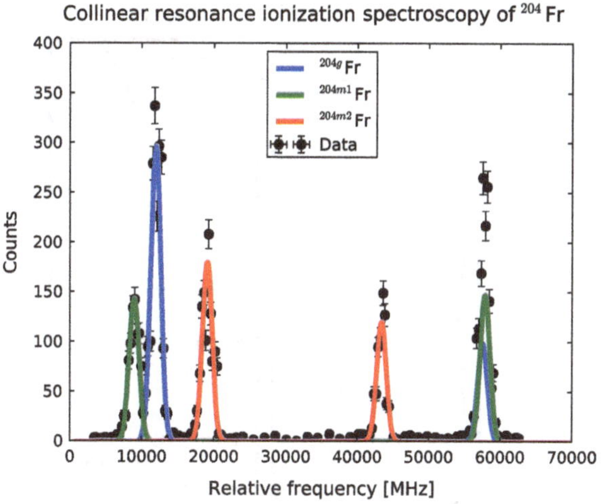

Fig. 7.16 Collinear resonance ionization spectroscopy scan of ^{204}Fr

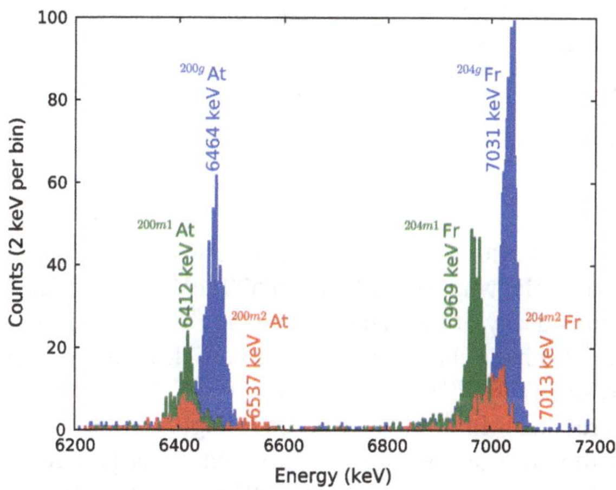

Fig. 7.17 Laser assisted nuclear decay spectroscopy of ^{204}Fr. The alpha-particle energy spectrum for ^{204g}Fr, ^{204m1}Fr and ^{204m2}Fr when the laser was set on resonance with a characteristic atomic transition in the hyperfine spectrum of each state

The alpha-particle energy spectrum of the laser assisted nuclear decay spectroscopy of ^{204g,m1,m2}Fr is presented in Fig. 7.17. The blue data shows the energy of the alpha particles emitted when the laser was on resonance with an atomic transition of the hyperfine spectrum characteristic of ^{204g}Fr. This transition occurred at 11837.54 cm^{-1}. Similarly, the green and red data show the alpha-particle energy spectrum for ^{204m1}Fr and ^{204m2}Fr, when the laser was tuned to 11837.49 cm^{-1} and 11837.66 cm^{-1}.

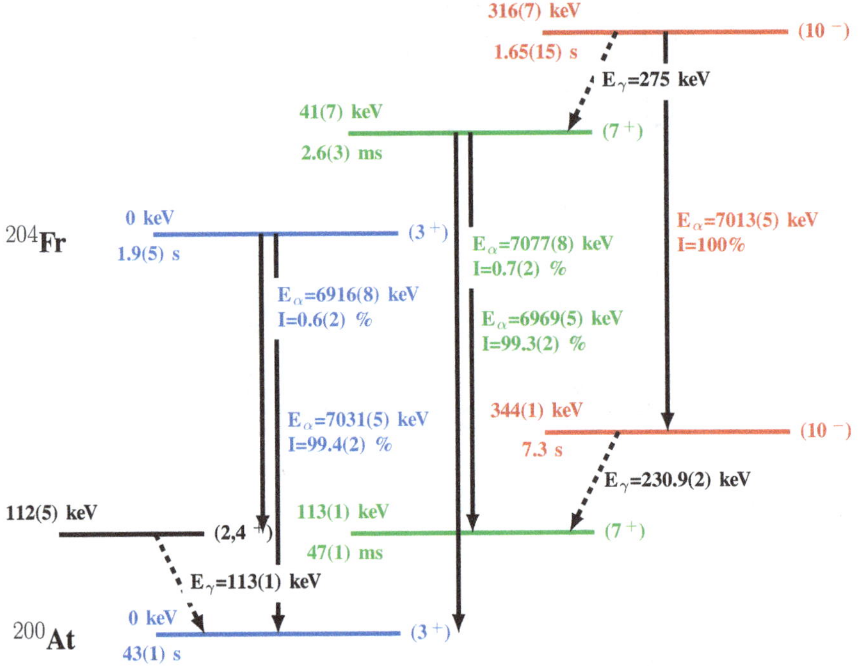

Fig. 7.18 The radioactive decay of ^{204}Fr and its isomers [9–12]

The alpha particles with energies approximately 7,000 keV originate from the decay of the state in ^{204}Fr, whereas the 6,400–6,500 keV alpha particles are associated with the radioactive decay of the daughter isotope ^{200}At. From Fig. 7.17 it is clear that each state in ^{204}Fr has a characteristic alpha-particle emission energy: 7,031 keV for ^{204g}Fr, 6,969 keV for ^{204m1}Fr and 7,013 keV for ^{204m2}Fr. The radioactive decay scheme of the three states in ^{204}Fr can be seen in Fig. 7.18.

Due to the difference of energy of the emitted alpha particles for ^{204g}Fr and ^{204m2}Fr of 18 keV, an energy resolution of $\leq$18 keV for the APIPS detector would be needed in order to sufficient resolve these two peaks [13]. Previous alpha-decay spectroscopy performed on ^{204}Fr achieved an energy resolution of 11–17 keV at 5.486 MeV with silicon PIPS detectors (50–150 mm^2) and 20 keV with surface barrier detectors (450 mm^2) [9]. This resulted in the ability to resolve the energies of the alpha particle emitted by ^{204g}Fr and ^{204m1}Fr for the first time. Despite a manufacture's specification of 19 keV at 5.486 MeV for the APIPS detector, only 30 keV at 6.341 MeV was achieved during the experimental run, as discussed in Sect. 6.3.5. This meant that without the ability to purify the ion beam with the CRIS technique, the identification of the two alpha-decay peaks would not be possible.

An additional 20 min radioactive decay measurement was performed on the fourth peak in the hyperfine spectrum of ^{204}Fr (see Fig. 7.16) at 11838.07 cm^{-1}. The observation of alpha particles with an energy of 7,013 keV allowed this state to be identified

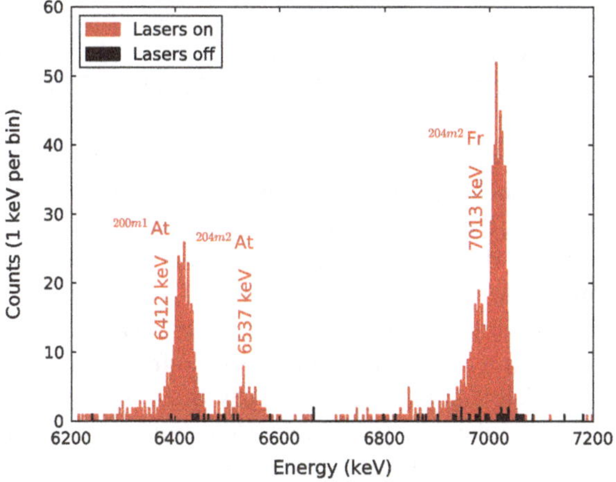

Fig. 7.19 Lasers on/off measurement of ^{204m2}Fr

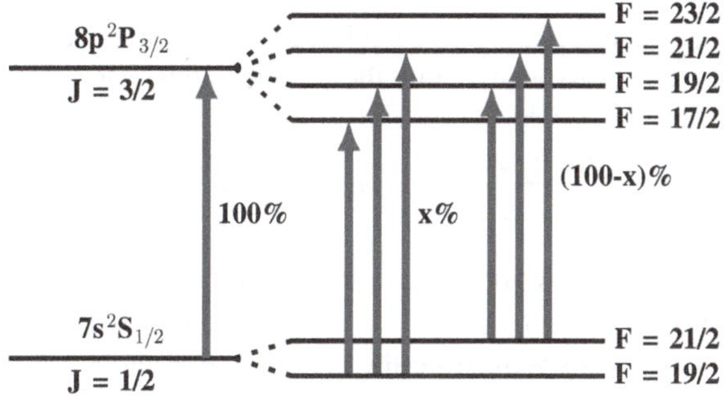

Fig. 7.20 The hyperfine splitting of ^{204m2}Fr. The addition of the alpha spectra when the laser was on resonance with each of the two transitions (shown in the diagram as x % and (100−x) %) gives the total atomic transition (100 %)

as ^{204m2}Fr. This meant the identity of all five hyperfine structure peaks could be allocated to a state in ^{204}Fr (by process of elimination the fifth and final hyperfine structure peak must be the overlapping structure of ^{204g}Fr and ^{204m1}Fr), and the analysis of the hyperfine structure of each state could be undertaken. This is discussed in detail in Sect. 7.2.2.

Figure 7.19(red) shows the combined alpha-particle energy spectrum when laser assisted nuclear decay spectroscopy was performed on the two hyperfine structure peaks of ^{204m2}Fr. This was at a wavenumber of 11837.66 cm^{-1} and 11838.07 cm^{-1} respectively. A schematic of these two transitions can be seen in Fig. 7.20. The addition of the two peaks (one of x %, the other with (100−x) %) gives the total (100 %) atomic transition for the ^{204m2}Fr isomeric state.

For each decay spectroscopy measurement of 20 min performed on the two hyperfine peaks, an additional 20 min measurement was made, where the laser beam for the resonant step (422.7 nm) was blocked. The non-resonant step (1,064 nm) was still transmitted through the beam line, to avoid modifying the DGF parameters of the digital DAQ system and to keep the rest of the experimental conditions constant. The alpha particles detected during these measurements are shown in Fig. 7.19(black). These alpha particles are the result of non-resonant collisional ionization of the francium isotopes with gas molecules in the interaction region. A small number of alpha particles detected may be the result of part of the ion beam being implanted into collimator (before the APIPS detector), with the decay of this activity via alpha-emission detected by the other side of the APIPS detector.

The total number of alpha particles detected when the lasers were on ($N(\alpha)_{\text{on}}$) and when they were off ($N(\alpha)_{\text{off}}$) was integrated over the francium and astatine energy range. The alpha particle emitted from the decay of astatine were included in the measurement to increase the statistics of the calculation and make it more precise. The ratio of laser-on to lasers-off was determined to be

$$\frac{N(\alpha)_{\text{on}}}{N(\alpha)_{\text{off}}} = 103^{+27}_{-18}.$$

The asymmetry of the error bars was calculated from a upper and lower estimate of the ratio from the errors associated with the detected alpha particles

$$\text{Upper limit} = \frac{N(\alpha)_{\text{on}} + \sqrt{N(\alpha)_{\text{on}}}}{N(\alpha)_{\text{off}} - \sqrt{N(\alpha)_{\text{off}}}}, \text{ Lower limit} = \frac{N(\alpha)_{\text{on}} - \sqrt{N(\alpha)_{\text{on}}}}{N(\alpha)_{\text{off}} + \sqrt{N(\alpha)_{\text{off}}}}.$$

The increase in alpha particles detected when the lasers were on resonance with the atomic transitions was at least 10^2. However, based on the assumption that many of the $N(\alpha)_{\text{off}}$ alpha particles were from the ^{221}Fr activity implanted into the APIPS collimator, a better focusing of the ion beam through the collimator and APIPS aperture is expected to increase the ratio significantly.

An additional collection of 60 mins was performed for ^{204m2}Fr. Figure 7.21 shows the alpha-particle energy spectrum of the combined statistics for this isomer. This spectrum illustrates the presence of the 6,969 keV alpha particle (denoted by $\star$), emitted from the decay of ^{204m1}Fr. The decay of ^{204m2}Fr to ^{204m1}Fr via an E3 internal transition (IT) has been predicted [9] but only recently observed [14]. This was achieved by tagging the conversion electron from the internal conversion of ^{204m2}Fr with the 6,969 keV emitted alpha particles of ^{204m1}Fr that followed (with a 5 s correlation time). This allowed the predicted energy of the 275 keV isomeric transition to be determined.

The pure isomeric beam of ^{204m2}Fr produced with the CRIS technique allows the 6,969 keV alpha particles from the decay of ^{204m1}Fr (via the E3 IT of ^{204m2}Fr) to be observed. These alpha particles can be clearly seen in the alpha-particle energy spectrum of Fig. 7.21. However, the conversion electrons from the internal transition of ^{204m2}Fr were not observed due to the high threshold set for the silicon detectors (a result of the high noise on the signals due to the 1,064 nm laser light). This is further discussed in Sect. 8.4.2.

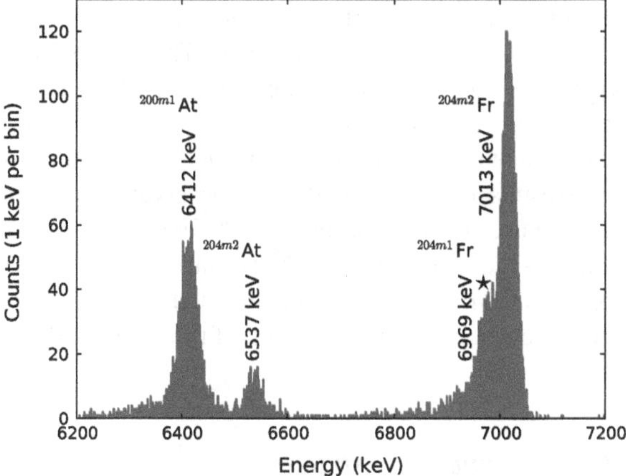

Fig. 7.21 Alpha-particle energy spectrum of ^{204m2}Fr. The presence of the 6,969 keV alpha particle emitted from the E3 IT to ^{204m1}Fr is denoted by ⋆

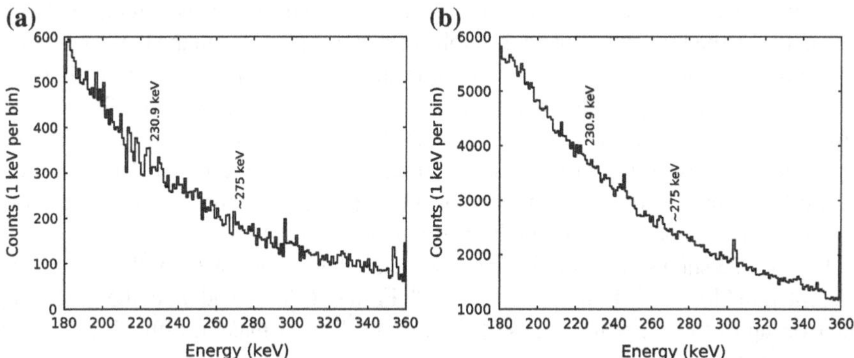

Fig. 7.22 Gamma-ray energy spectrum of ^{204m2}Fr from **a** ΔE crystal of the E-ΔE detector, and **b** the HPGe detector. Neither the 230.9 keV gamma ray from the decay of ^{200m2}At nor the 275 keV line from the decay of ^{204m2}Fr are observed. This is attributed to the low detection efficiency of the germanium detectors (see Sect. 6.3.1)

The presence of the 275 keV gamma ray from the E3 IT from ^{204m2}Fr to ^{204m1}Fr was investigated in the energy spectra measured by the germanium detectors. Figure 7.22 shows the gamma-ray energy spectra from the (a) ΔE crystal of the E-ΔE detector, and (b) the HPGe detector. In both of these spectra, the 275 keV gamma line is not present. However, the established 230.9 keV (from the decay of ^{200m2}At to ^{200m1}At via gamma-ray emission) is not observed either. The absence of this gamma-ray line is attributed to the very low detection efficiency of the germanium detectors surrounding the DSS. A more detailed discussion of this can be found in Sect. 6.3.1.

The alpha branching ratio of ^{204m2}Fr was calculated by comparing the relative intensities of the mother and daughter alpha decays. The value was determined to be 53(10) %, in disagreement with the branching ratio of 74(8) % from literature [15]. This measurement compared the observation of the ^{204m2}Fr and ^{200m2}At alpha particles with the emission of known gamma rays in ^{204}Rn. In addition, the decay of ^{200m2}At allowed the results to be confirmed by the observation of the gamma decay in ^{200}Po. The CRIS result used only the observed ^{204m2}Fr alpha decay in comparison to the population and subsequent decay of ^{200m2}At, introducing inaccuracy. The main source of uncertainty in the calculation was associated with the observed number of alpha particles due to low statistics in the fitting of the alpha peaks.

7.2.2 Alpha-Gated Hyperfine Structure of ^{204}Fr

Just as the laser frequency of the resonant 422.7 nm ionization step was scanned and resonant ions were detected by the MCP in the collinear resonance ionization spectroscopy of ^{204}Fr, the same technique was repeated with the measurement of alpha particles. At each laser frequency, a radioactive decay measurement of 1 min was made by the DSS, measuring the alpha particles emitted from the ions implanted at that time. Figure 7.23 shows the hyperfine structure of ^{204}Fr measured with alpha particles. The alpha-particle energy gate was set between 6,200 and 7,200 keV in order to measure both the francium and astatine alpha particles emitted from the implanted ions, resulting in increased statistics to allow for a better determination of the hyperfine factors. The number of alpha particles in this energy range was integrated and associated with the laser frequency at that time.

Figure 7.23 shows the hyperfine peak associated with each state in ^{204}Fr: ^{204g}Fr is presented in blue, ^{204m1}Fr in green, and ^{204m2}Fr in red. As is evident in the spectrum, the overlap of the hyperfine peaks of ^{204g}Fr and ^{204m1}Fr make it difficult to extract accurate values of the hyperfine parameters due to the error in the location of the two right-hand peaks of the states in the peak envelope. Figure 7.24 shows a two dimensional representation of the hyperfine structure of ^{204}Fr with alpha-particle energy. As the frequency is scanned, the energy of the emitted alpha particles changes as each isomer is resonantly ionized and its decay detected.

In order to extract clean hyperfine structures for each states, an alpha-energy gate was applied to the detected alpha-particles. The ground state of ^{204}Fr emits a characteristic alpha particle of energy 7,031 keV, so an energy gate of 7,031–7,200 keV was applied. In a similar manner, the ^{204m1}Fr alpha particles were identified by an energy gate of 6,959–6,979 keV and for the ^{204m2}Fr alpha particles, an energy gate of 7,003–7,023 keV was applied. As the emitted alpha particles of ^{204g}Fr and ^{204m2}Fr have similar energies (7,031 and 7,013 keV respectively) the energy gates were chosen to maximise the number of the alpha particle of interest, while limiting the number of alpha particles associated with the isomers that were not of interest.

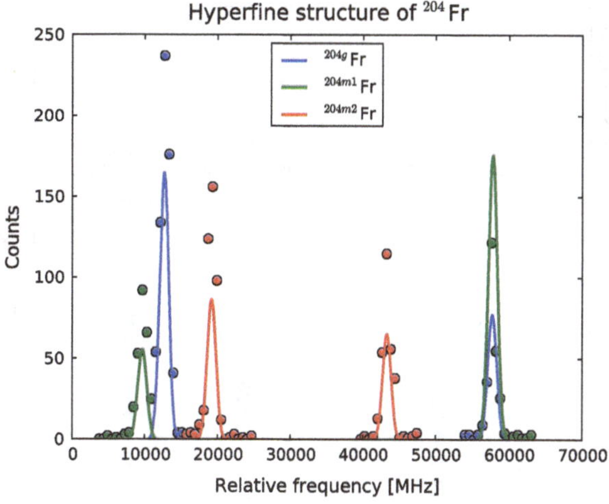

Fig. 7.23 Hyperfine structure of ^{204}Fr measured by the DSS. The hyperfine structure of ^{204}Fr was scanned and the integrated number of alpha particles per frequency step is plotted

Table 7.8 Extracted hyperfine observables from the alpha-gated hyperfine spectra of ^{204}Fr

Isotope	Spin	$A_{S_{1/2}}$ (MHz)	Isotope shift (MHz)
204g	(3^+)	+12990(30)	32190(100)
204m1	(7^+)	+6440(30)	32320(100)
204m2	(10^-)	+2310(30)	30990(100)

By gating on characteristic alpha-particle energies of the three states in ^{204}Fr, clean hyperfine structures can be obtained, as shown in Fig. 7.25. Figure 7.25a shows the hyperfine structure of ^{204g}Fr, (b) ^{204m1}Fr, and (c) ^{204m2}Fr. The presence of ^{204m2}Fr can be observed in the spectra of ^{204g}Fr and ^{204m1}Fr, due to the overlapping peaks of the alpha energies: the tail of the 7,013 keV alpha peak is present in the gate of the ^{204g}Fr and ^{204m1}Fr alpha peaks. In a similar manner, ^{204g}Fr is present in the ^{204m2}Fr spectra due to the similar energies of the 7,031 and 7,013 keV alpha particles. However, despite the contamination in the hyperfine spectra, each peak is separated sufficiently in frequency to cleanly fit.

From the resulting clean hyperfine structures given by alpha particles, in comparison to the overlapping ion data of Fig. 7.16, each state of ^{204}Fr can be analysed individually and the hyperfine factors extracted with a larger degree of accuracy. Table 7.8 presents the $A_{S_{1/2}}$ factors and isotope shifts of the three states in ^{204}Fr. The error attributed to the $A_{S_{1/2}}$ factors was 30 MHz on account of the scatter of $A_{S_{1/2}}$ values for ^{221}Fr. Likewise, an error of 100 MHz was assigned to the isotope shifts.

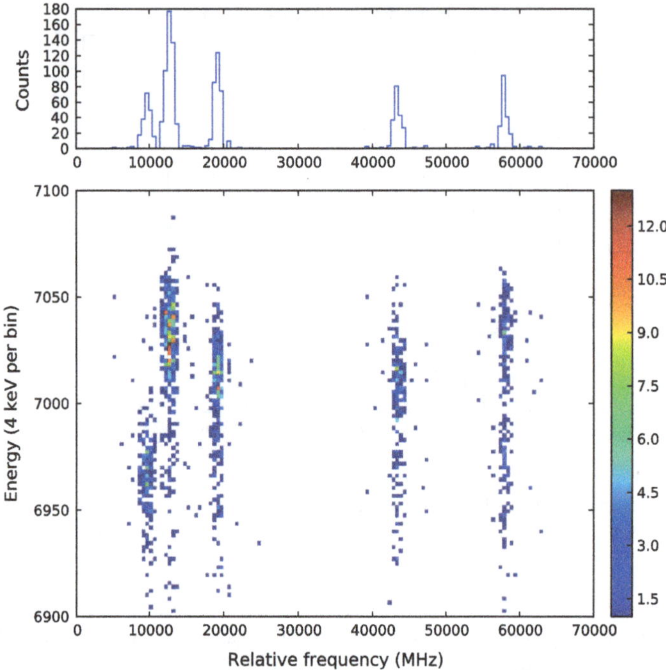

Fig. 7.24 *Below* a 2D histogram of the alpha-particle energy as the hyperfine structure is scanned. *Above* projection of the x-axis. The total number of alpha particle detected at each laser frequency reveals the hyperfine structure of ^{204}Fr

7.2.3 Laser Assisted Nuclear Decay Spectroscopy of ^{202}Fr

In a similar manner to ^{204}Fr, laser assisted nuclear decay spectroscopy was performed on the ground and isomeric state of ^{202}Fr. The laser was tuned onto resonance with the first (ground state) and second (isomeric state) peak in the hyperfine spectrum shown in Fig. 7.26 obtained from ion detection. With the laser on each resonant transition (11837.62 cm^{-1} and 11837.74 cm^{-1} respectively), a decay measurement of 15 mins was performed. The alpha-particle energy spectrum is shown in Fig. 7.27. The alpha particles emitted when the laser was on resonance with an atomic transition characteristic to ^{202g}Fr are shown in blue, and ^{202m}Fr in red.

Evident in the spectrum of ^{202g}Fr are the alpha particles emitted from the decay of the daughter nucleus ^{198g}At with an energy of 6,755 keV. Similarly, present in the ^{202m}Fr spectrum are the alpha particles from the decay of ^{198m}At with an energy of 6,856 keV. The difference in energy of these two alpha peaks illustrates the ability of the CRIS technique to separate the two states and provide pure ground state and isomeric beams for decay spectroscopy.

According to literature, the radioactive decay of ^{202g}Fr ($t_{1/2}$ = 0.30(5) s) emits an alpha particle of energy 7,241(8) keV, whereby ^{202m}Fr ($t_{1/2}$ = 0.29(5) s) emits

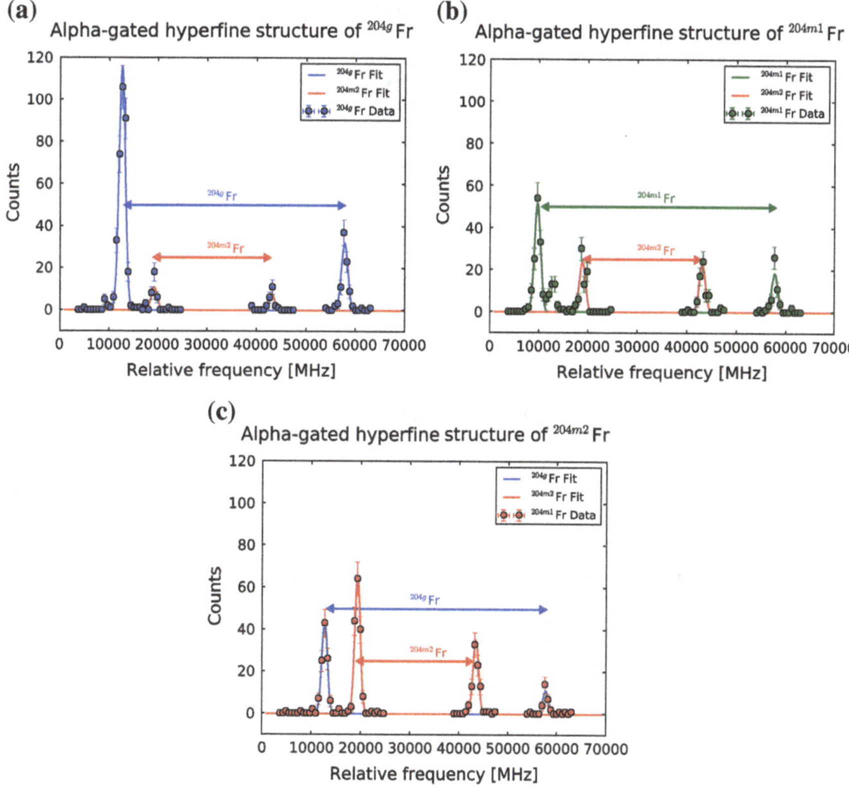

Fig. 7.25 Gating on the characteristic alpha particle energies (7031, 6969 and 7013 keV) yields clean hyperfine structures for **a** ^{204g}Fr, **b** ^{204m1}Fr and **c** ^{204m2}Fr respectively

an alpha particle of energy 7,235(8) keV [16]. Due to the limited statistics of our measurement, and the similarity in energies of the alpha particles (within error), it is impossible to say that alpha particles of different energies are observed in Fig. 7.27. The mean energy of the alphas emitted by ^{202g}Fr and ^{202m}Fr differ by 8 keV, which implies that the alpha particles in the two spectra have slightly different energies. However, limited statistics from this experiment means that no conclusions on the nature of these particles can occur. A repeat measurement with significantly higher statistics of laser assisted nuclear decay spectroscopy of ^{202g,m}Fr would provide detailed knowledge of the alpha-particle energies and half-lives of the two states. Assuming the same experimental efficiency, 1.5 shifts (of 8 h each) would allow sufficient statistics to be obtained. This would give over 1,000 integrated counts for the ^{202g}Fr peak (>100 counts at the peak energy 7,241(8) keV) and ~600 integrated counts for ^{202m}Fr (assuming identical production intensities).

The mean energy of the alpha particles emitted by ^{202g}Fr and ^{202m}Fr was determined to be 7,267 keV and 7,259 keV respectively. The difference in the detected energy with the literature values (7,241(8) keV and 7,235(8) keV) is a result of a drift

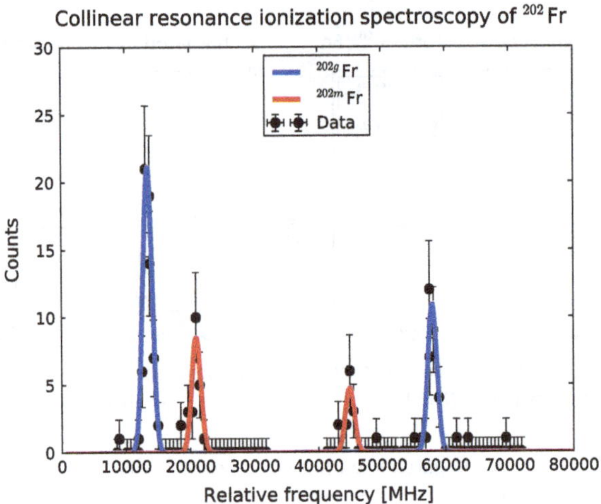

Fig. 7.26 Collinear resonance ionization spectroscopy scan of ^{202}Fr

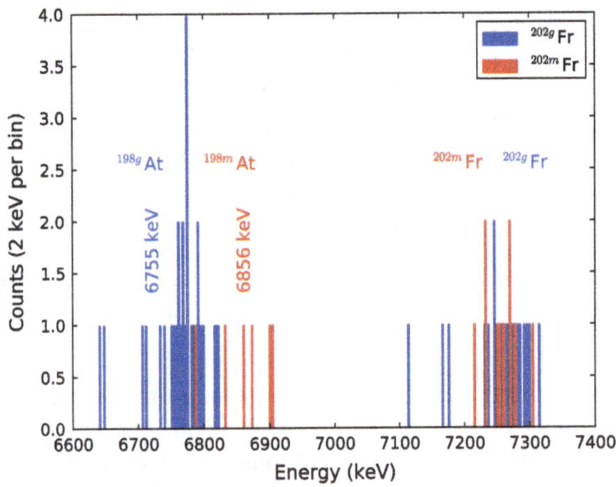

Fig. 7.27 Laser assisted nuclear decay spectroscopy of ^{202}Fr. The alpha spectrum for ^{202g}Fr and ^{202m}Fr when the laser was set on resonance with a characteristic atomic transition in the hyperfine spectrum of each state

in the calibration of the APIPS silicon detector. This drift was mainly due to the variation in the silicon detector's baseline associated with the 1,064 nm laser light, as discussed in Sect. 6.2.3. Since no reference scan of ^{221}Fr was taken close in time, the laser assisted decay spectroscopy of ^{202}Fr was carried out without calibration of the silicon detector with alpha particles of known energies.

The total experimental efficiency of the CRIS technique is a result of the combined efficiencies of transmission, detection, laser ionization and neutralization,

$$\epsilon_{\text{experimental}} = \epsilon_{\text{transmission}}\, \epsilon_{\text{detection}}\, \epsilon_{\text{laser ionization}}\, \epsilon_{\text{neutralization}}. \tag{7.6}$$

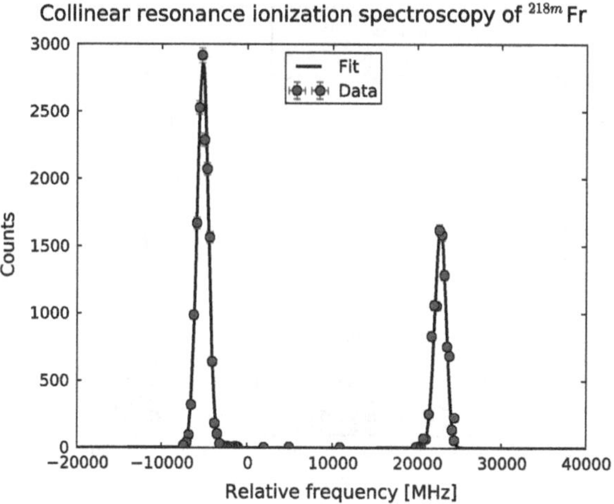

Fig. 7.28 Collinear resonance ionization spectroscopy scan of ^{218m}Fr

The yield of ^{202}Fr was independently measured after ISCOOL by an alpha-spectroscopy decay station. For every 130 alpha particles emitted from ^{202}Fr that was observed at the decay setup, one resonant ion was detected with the MCP. As the decay station measured the total yield of ^{202}Fr (ground state and isomer) the laser ionization efficiency of ionizing either ground or isomeric state will be higher. Ignoring the population of each of the 7s $^{2}S_{1/2}$ states (50 % each), a conservative estimate of the total experimental efficiency of 1 % was determined.

7.2.4 Laser Assisted Nuclear Decay Spectroscopy of ^{218m}Fr

The half-life of ^{218g}Fr is 1.0(6) ms resulting in the isotope decaying before it can be transported to the CRIS setup. However, the longer-lived isomeric state of ^{218m}Fr, 86(8) keV above the ground state ($t_{1/2}$ = 22.0(5) ms) was observed [17]. The hyperfine spectrum of ^{218m}Fr can be seen in Fig. 7.28. The hyperfine structure studies of ^{218m}Fr, along with the other neutron-rich francium isotopes of ^{219,229,231}Fr measured during the experimental campaign are in preparation for publication and are discussed in the Ph.D. thesis of Budinčević [18].

The laser frequency was tuned onto resonance with each atomic transition of the hyperfine spectrum of ^{218m}Fr, shown in Fig. 7.28. The alpha-particle energy spectrum collected during these measurements is presented in Fig. 7.29. The blue spectrum, obtained when the laser light was tuned to 11836.98 cm^{-1}, shows the alpha particles emitted when the laser was on resonance with the left-hand peak in the hyperfine spectrum. Similarly, the green spectrum in Fig. 7.29 shows the alpha particles emitted when the laser light had a wavenumber of 11837.435 cm^{-1}, on resonance with the right-hand peak of Fig. 7.28. Present in the two spectra are the alpha particles emitted

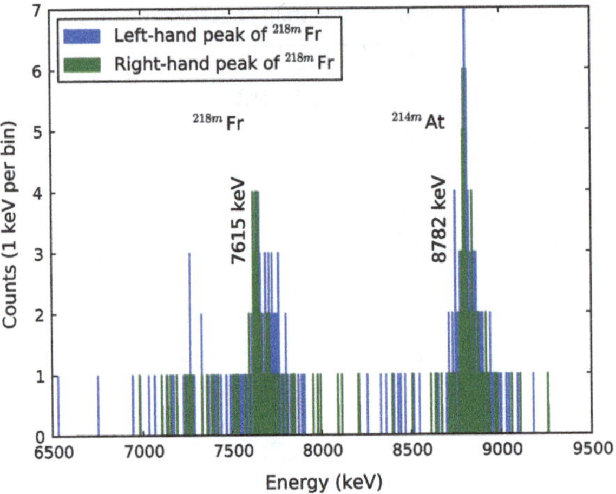

Fig. 7.29 Laser assisted nuclear decay spectroscopy of ^{218m}Fr

from the decay of ^{218m}Fr (7,615 keV) and the daughter nucleus ^{214m}At (8,782 keV). The increase in the number of alpha particles in the blue spectrum, in comparison to the green, is due to the intensities of the hyperfine structure peaks. Integrating over the complete energy range (to increase statistics by including the daughter alpha particles), the ratio is

$$\text{Ratio}_{\alpha} = \frac{N(\alpha)_{11836.98}}{N(\alpha)_{11837.435}} = 2.0(2).$$

In comparison with the CRIS ion data, the ratio of the peak intensities of the two hyperfine structure peaks is

$$\text{Ratio}_{\text{ion}} = \frac{N(\text{ion})_{11836.98}}{N(\text{ion})_{11837.435}} = 1.8(1),$$

in excellent agreement with the ratio of the alpha data.

In addition to the ratio of the hyperfine peak intensities, the ratio of laser-on to laser-off was calculated. A 15 min radioactive decay measurement was performed when the resonant 422.7 nm laser light was blocked from entering the interaction region. During this time, no alpha particles in the energy range 6,500 to 9,500 keV were observed. This represented the non-resonant ionization efficiency of the setup, and is given by

$$\epsilon_{\text{non-resonant}} = \epsilon_{\text{transmission}}\, \epsilon_{\text{detection}}\, \epsilon_{\text{non-resonant ionization}}\, \epsilon_{\text{neutralization}}. \tag{7.7}$$

The number of alpha particles observed in Fig. 7.29 was integrated over the same energy range. Following the same procedure as outlined in Sect. 7.2.1, the laser-on

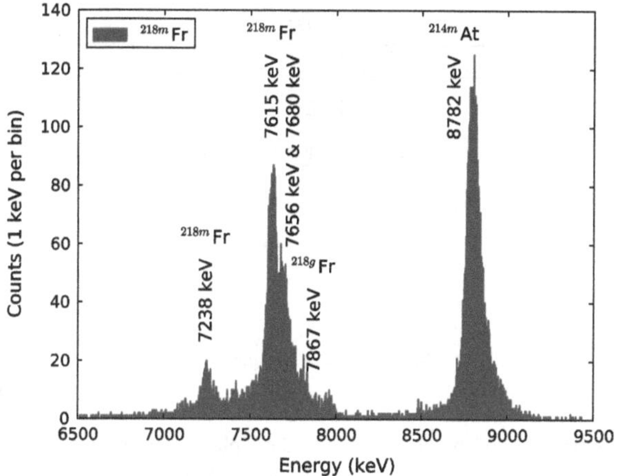

Fig. 7.30 Alpha-particle energy spectrum of ^{218m}Fr

to laser-off ratio was calculated as

$$\text{Ratio} = \frac{N(\alpha)_{\text{on}}}{N(\alpha)_{\text{off}}} >> 334.$$

To avoid a null result, $N(\alpha)_{\text{off}}$ was modified from 0 counts to 1(1). This allowed a lower limit on the ratio to be calculated. As no alpha particles were observed when the laser light was blocked, the ratio of lasers-on to laser-off can be considered to be much greater than the presented value of 334. However, this value illustrates the increase in experimental efficiency that took place during the experiment: the lasers-on to lasers-off ratio has increased significantly since the ratio of 103^{+27}_{-18} for ^{204}Fr (see Sect. 7.2.1).

Combining Eqs. 7.6 and 7.7, the laser-on to laser-off ratio can be considered to be

$$\text{Ratio} = \frac{\epsilon_{\text{laser ionization}}}{\epsilon_{\text{non-resonant ionization}}}.$$

Combining the 30 % laser ionization efficiency (calculated from the total experimental efficiency) and a non-resonant ionization efficiency of 1:10^5 (calculated from the pressure of the interaction region and the cross section of ionization), the ratio is 1 in 3×10^4. This theoretical value is significantly higher than the experimental lower limit of 334, and highlights the efficiency of the resonant ionization process: a much longer collection time would have been needed in order to observe an alpha particle from the non-resonant ionization of ^{218}Fr. The experimental efficiency of the CRIS experiment is discussed in detail in the Ph.D. thesis of T. J. Procter [19].

Figure 7.30 shows the energy of the alpha particles emitted when the total beam of ^{218m}Fr produced by ISOLDE was implanted into a carbon foil. This was achieved by setting the voltage on the deflector plate (which deflects the non-neutralized

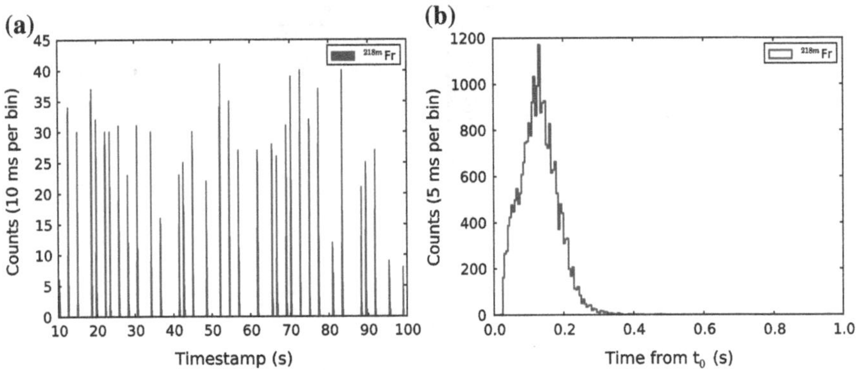

Fig. 7.31 The time structure of the observation of the alpha-decay of ^{218m}Fr. **a** The timestamp of the alpha particles **b** The saturation and decay activity curve of ^{218m}Fr

component of the beam, situated after the CEC) to 0 V. A decay measurement of 5 mins was collected. As mentioned in Sect. 7.2.3, no calibration measurement was taken close in time to the implanting of the ^{218m}Fr beam, and thus no accurate calibration could be used. Despite this, it was possible to identify the alpha particles emitted from the decay of ^{218m}Fr and its daughter nuclei ^{214m}At. There is a slight peak around 7,900 keV which is unidentified, but due to the drift in the calibration of the spectra, could be due to the decay of the short-lived ^{218g}Fr.

Figure 7.31a shows the timestamp of the alpha particles detected during the measurement. The plot shows the time window 10–100 s to illustrate the time structure of the alpha particles. From the radioactive decay of ^{218m}Fr and ^{214m}At, the alpha particles are detected in pulses 1.2 s apart. This corresponds to the impact of protons on the ISOLDE target, also at 1.2 s intervals. The time of each proton impact (t_0) was set to zero and the alpha particle detection time (time from t_0) was plotted. The resulting saturation and decay activity curve is presented in Fig. 7.31b. This plot shows the alpha particles emitted from the decay of ^{218m}Fr and ^{214m}At: an alpha-energy gate of 6,500–9,500 keV was applied. Since the half-life of ^{214m}At is 760(15) ns, it can be considered virtually instantaneous. This results in the time at which the ^{214m}At alpha particles are detected being solely dependent on the half-life of the parent isotope ^{218m}Fr. From the impact of the proton, the number of alpha particles emitted from the decay of ^{218m}Fr increases until there is a saturation at $t_0 = 0.1$ s. Once the activity has saturated the implanted activity decays, allowing the half-life of the isomer to be measured.

The half life of a radioactive isotope can be calculated from the expression

$$N(t) = N_0 \exp -\lambda t, \tag{7.8}$$

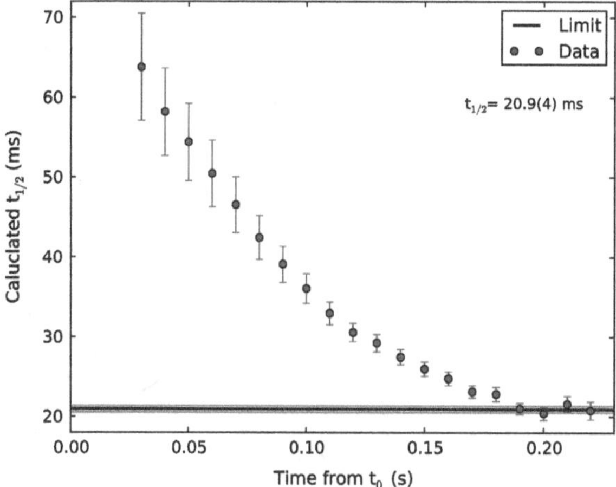

Fig. 7.32 Analysis of the half life of ^{218m}Fr. Convergence of half-life at $t_{1/2} = 20.9(4)$ ms is very close to agreement agreement with the literature value of 22.0(5) ms [20]

where

$$t_{1/2} = \frac{\ln 2}{\lambda}, \tag{7.9}$$

and N_0 is the initial number of radioactive species, λ is the decay constant, and $t_{1/2}$ is the half-life of the radioactive isomer. The fitting of Eq. 7.8 to the exponential region of Fig. 7.31b was performed. This yielded λ which, through use of Eq. 7.9, allowed the half-life of ^{218m}Fr to be calculated.

Due to the saturation of the activity of ^{218m}Fr in Fig. 7.31b, the time over which the fitting of Eq. 7.8 was performed affected the calculation of the half-life. This is associated with the release of the ISOLDE target, since not all of the ^{218m}Fr atoms arrive a $t_0 = 0$, but rather over a period of several ion bunches. In order to converge on the correct value of the half-life, this time was varied.

Figure 7.32 shows how the half-life depends on when the fit is started. The errors on the half-life values were propagated from the errors on the fit parameters. There is a decrease in the value of the half-life, and at $t = 0.2$ s the value starts to converge. The mean (black line) and error (grey region) of the half-life was calculated from the last four data points of Fig. 7.32.

The half-life was determined to be

$$t_{1/2}(^{218m}\text{Fr}) = 20.9(4)\text{ ms}.$$

This is very close to agreement with the literature value of 22.0(5) ms [20]. Figure 7.33 shows the number of alpha particles detected after the proton impact at t_0 (blue data

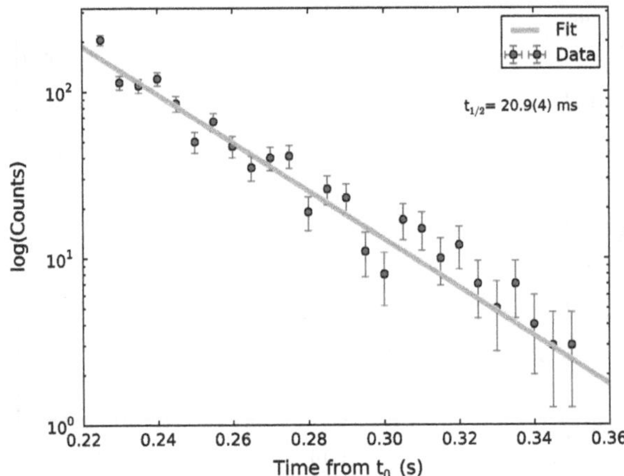

Fig. 7.33 Determination of the half life of ^{218m}Fr. Equation 7.8 was fit to the data, shown with a logarithmic y-axis

points). The red line is the fit to the data, using Eq. 7.8. The data is presented with a logarithmic y-axis.

References

1. Coc A et al (1985) Phys Lett B 163:66
2. Duong HT et al (1987) EPL 3:175
3. Blaum K et al (2013) Phys Scripta 2013:014017
4. Voss A et al (2013) Application of lasers and storage devices in atomic nuclei research. Poznan, Poland
5. Rothe S (2013) Private communication
6. Dzuba VA et al (2005) Phys Rev A 72:022503
7. Ekström C et al (1986) Phys Scripta 34:624
8. Gomez E et al (2008) Phys Rev Lett 100:172502
9. Huyse M et al (1992) Phys Rev C 46:1209
10. Uusitalo J et al (2005) Phys Rev C 71:024306
11. Jakobsson U et al (2013) Phys Rev C 87:054320
12. http://www.nndc.bnl.gov/nudat2/
13. Leo WR (1994) Techniques for nuclear and particle physics experiments. Springer, Berlin
14. Jakobsson U et al (2012) Phys Rev C 85:014309
15. Bingham CR et al (1995) ENAM95: International Conference on Exotic Nuclei and Atomic Masses p 545
16. Zhu S, Kondev F (2008) Nucl Data Sheets 109:699
17. Wu SC (2009) Nucl Data Sheets 110:681
18. Budinčević I (2014) Ph.D Thesis, IKS, KU Leuven
19. Procter (2013) TJ Ph.D Thesis, The Nuclear Group, University of Manchester
20. Ewan G et al (1982) Nucl Phys A 380:423

Chapter 8
Interpretation of Results

8.1 Charge Radii Across the N = 126 Shell Gap

The smooth change in the mean-square charge radii with mass number is associated with the pure volume effect at constant deformation and can be described by the droplet model, discussed in Sect. 3.2.1 [1, 2]. The evolution of the charge radii characterises the distribution of the protons in the nucleus, whereby the droplet model qualitatively describes $\delta\langle r^2\rangle$ as a function of Z and N. However, quantum shell effects can cause a divergence from the smooth systematics of the region. The gradient of the $\delta\langle r^2\rangle$ line doubles when the N = 126 shell gap is crossed in the lead isotopes. A similar effect can be seen in the other isotopes of this region, illustrated for the even-Z isotopes from lead (Z = 82) to radium (Z = 88) in Fig. 8.1. Theoretical models, such as Skyrme-Hartree-Fock or relativistic mean field (RMF) calculations, cannot consistently explain this observed 'kink'. Recent studies of the different theoretical models that do and do not reproduce this characteristic attribute the kink at N = 126 to the occupancy of the $\nu 1\mathrm{i}_{11/2}$ neutron shell [3]. With the addition of neutrons into the $\nu 1\mathrm{i}_{11/2}$ shell, the protons states are attracted to a larger radius, resulting in the ability to maximally overlap with the extra neutrons in the n = 1 orbital.

Figure 8.1 illustrates the systematics of the charge radii in the region of the Z = 82 shell closure. The ground state of the neutron-deficient lead isotopes (as the N = 126 closed shell is depleted) remains spherical, attributed to the robustness of the Z = 82 magic number [5, 6]. The mercury isotopes (Z = 80) display a strong deviation from sphericity around N = 104, the middle of the two closed shells of N = 82 and N = 126 [4], for the odd-A isotopes. As the nucleus becomes more neutron deficient, there is an increase in the polarizability of the nucleus by the single valence neutron in the $\nu \mathrm{p}_{1/2}$ orbital, and hence collectivity begins to play a role. The light even-A mercury isotopes display a small oblate deformation, where in contrast the light odd-A mercury isotopes are strongly prolate deformed. Below N = 106, the prolate shape of the nucleus gains 0.5 MeV in binding energy relative to the oblate shape. This is due to the competing near spherical and deformed bands. The odd-even staggering

K.M. Lynch, *Laser Assisted Nuclear Decay Spectroscopy*,
Springer Theses, DOI: 10.1007/978-3-319-07112-1_8

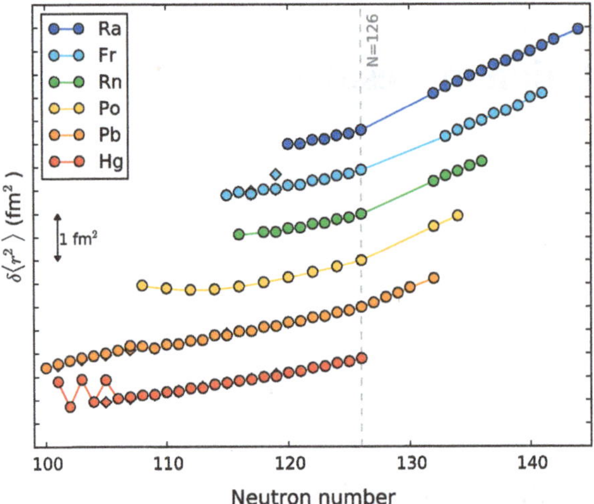

Fig. 8.1 Charge radii of nuclei in the Z = 82, N = 126 region of the nuclear chart: Hg (Z = 80) [4], Pb (Z = 82) [5–7], Po (Z = 84) [8], Rn (Z = 86) [9], Fr (Z = 87) [CRIS] [10] and Ra (Z = 88) [11]

observed is attributed to the transitions between different nuclear shapes, and the large isomer shifts indicate the presence of shape coexistence [4].

Due to their symmetry with the mercury isotopes across the Z = 82 shell closure, a similar trend from sphericity is expected to occur in the polonium (Z = 84) isotopes, with two protons outside the closed Z = 82 shell. However, a large deviation from sphericity occurs at N = 114, as shown in Fig. 8.2. The reasons for this early onset of deformation is the shift in the dominant configuration of the shape of the nucleus from spherical to oblate. This results from the change in location of the minima (corresponding to the ground state of the nucleus) in deformation energy space from near-spherical to oblate [8]. Results from beyond-mean field calculations suggests that there is a co-existence of different nuclear shapes at low excitation energies, leading to a softness in the neutron-deficient polonium isotopes.

The presence of strong octupole deformation is predicted in the Ra-Th region (Z = 88 to 90) at N = 136 [12]. This results from the coupling between states of opposite parity due to the long-range octupole-octupole residual interaction [7], where the interaction is strong enough to stably deform the nucleus in its rest frame. This onset of reflection asymmetry occurs when the intruder sub-shell (l, j) and normal parity sub-shell $(l-3, j-3)$ have a maximum $\Delta N = 1$ interaction. This can be seen just above the closed shells of 34, 56, 88 and 134. In the Ra-Th region, this occurs at N = 136 from the coupling of $\nu j_{15/2}$ and $\nu g_{9/2}$.

The radon (Z = 86) isotopes border the region of reflection asymmetry, localized in the Ra-Th chains. The odd-even staggering in isotope shift is inverted at N = 133 and 135, which is attributed to the presence of strong octupole correlations that are thought to develop into a static deformation of the odd-A isotopes in radon. This is due to the polarization of unpaired neutrons on the nucleus [9]. The light radium (Z = 88)

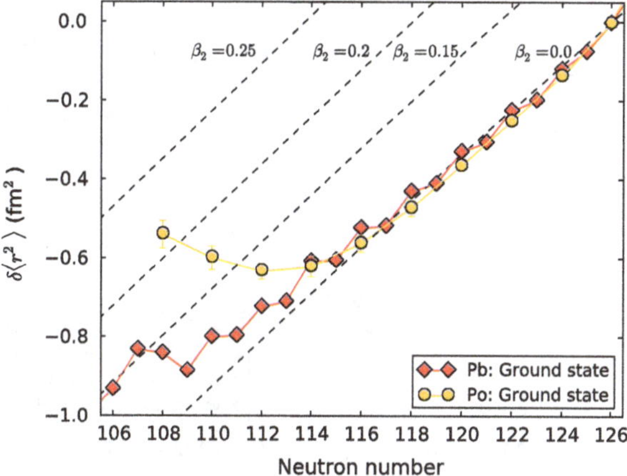

Fig. 8.2 Charge radii of even-A Po (Z = 84) [8] and Pb (Z = 82) [5] isotopes

isotopes display near-spherical shapes, where the small deviation from sphericity is due to the increasing dynamic collective motion on both sides of the magic neutron number. In the heavier radium isotopes, the transitional region of N = 130 to 136 marks the introduction of octupole deformed shapes. The recent observation of octupole deformation in ^{224}Ra (N = 136), and the weaker octupole collectivity of ^{220}Rn (N = 134), from electric octupole (E3) transition strengths suggest a static octupole deformation in the intrinsic frame of the nucleus. For the radium isotopes, the odd-even staggering is inverted for N = 131 and 133. The error on the charge radii comes from the uncertainty on the theoretical field shift factor, F. However, recent *ab initio* many-body calculations of the field shifts and specific mass shifts for the radium isotopes have extended the previous work in this area [11].

Located between radon and radium, francium (Z = 87) has an odd proton occupying the $\pi\, 1h_{9/2}$ orbital. Similar to its neighbours, the heavy francium isotopes exhibit strong octupole deformation at N = 136, which is discussed in detail in Sect. 8.2. Below the N = 126 shell closure, the neutron-deficient francium isotopes were studied down to ^{202}Fr (N = 115). The change in mean-square charge radii for the francium and lead isotopes are presented in Fig. 8.3. The data of francium show the Dzuba [10] charge radii of $^{207-213}$Fr alongside the CRIS values which extends the data set to ^{202}Fr. The blue data points show the francium ground states, with the 7^+ states in green and 10^- states in red. The error bars attributed to the CRIS values are propagated from the experimental error of the isotope shift and the systematic error associated with the atomic factors F_{422} and M_{422}. It is the systematic error that is the significant contribution to the error on the mean-square charge radii, and not that of the isotope shift. The francium data is presented with the lead data of Anselment [5] to illustrate the departure from the spherical nucleus. It is worth noting at this point that the charge radii of the lead isotopes have been overlapped with the charge radii of the francium isotopes. The charge radii of ^{208}Pb, conventionally determined to be

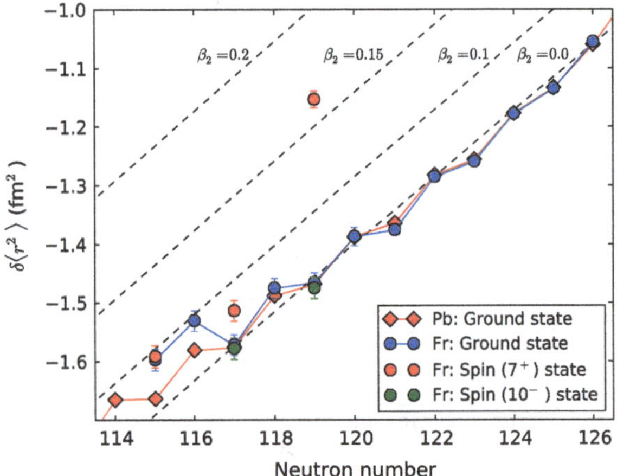

Fig. 8.3 Charge radii of Fr (Z = 87) [CRIS] [10] and Pb (Z = 82) [5] isotopes

$\delta\langle r^2\rangle = 0.0\,\text{fm}^2$, has been calibrated to equal the charge radii of ^{213}Fr (at N = 126) of $\delta\langle r^2\rangle = -1.06(1)\,\text{fm}^2$.

The doubly-magic ^{208}Pb (N = 126) represents a model spherical nucleus, with the sphericity of the nucleus remaining constant with the removal of neutrons from the closed N = 126 shell. The change in mean-square charge radii for the francium isotopes shows agreement with the lead data as the $\nu 3p_{1/2}$ and $\nu 2f_{5/2}$ orbitals are depleted. The deviation from sphericity at N = 116 with ^{203}Fr marks the onset of deformation. Without determination of the electronic quadrupole moment provided by high resolution laser spectroscopy measurements, it is not possible to say to what extent this change in mean-square charge radii by ^{203}Fr and below is due to static or dynamic deformation. Measurement of Q_s for the neutron-deficient francium isotopes will provide the time-averaged static deformation of the nucleus, determining the nature of this deformation.

Recent measurements at TRIUMF [13] of ^{206}Fr and its isomers have shown the ground state of ^{206}Fr to be consistent with the mean-square charge radii of the lead isotope ^{201}Pb. As seen in Fig. 8.3, ^{206g}Fr (N = 119) overlaps the lead data within errors. The small deviation from the lead data can be attributed to the calibration of the ^{206}Fr data from August to the October data, as only one hyperfine structure scan of ^{206}Fr taken during the experimental run in August. Irrespective of the small error that may be present in the charge radii due to the calibration, the large change in mean-square charge radius of ^{206m2}Fr suggests a highly deformed state for the isomer.

8.2 Odd-Even Staggering

The change in the mean-square charge radii with the addition of one unpaired neutron compared to the effect of adding a neutron pair can be illustrated by the odd-even staggering (OES) parameter of Tomlinson and Stroke [14]. The addition of a pair of neutrons results in isotope shifts (and by definition charge radii) that are very nearly equal. In contrast, the addition of an odd neutron to a even-N system will result in less than half the shift created by the pair of neutrons. This means the centroid frequency of the hyperfine structure of an odd-N isotopes will be closer to the lighter even-N isotope than its heavier even-N neighbour. This results from the competition between pairing energy and the single particle energy. Pairs of nucleons favour higher angular momentum states, where odd neutrons favour lower angular momentum states as no pairing energy is possible. This results in the decrease in effectiveness of the single valence neutron at producing distortions in the nucleon distribution in the nucleus.

The physical basis of OES, present across the nuclear chart, is still not fully understood. Talmi [15] attributed it to the polarization of the core by valence neutrons, whereby paired and unpaired neutrons resulted in different polarizations. Zawischa [16] explained the phenomena by four-particle (or alpha-) clustering, producing strong couplings between the neutron and proton pairing correlations, producing the staggering. The proton pairing results in an increase in the charge radius of the system due to the particle occupancy of the proton levels above the Fermi surface. As the higher energy levels generally have a larger radius, this results in larger radii for the proton paired nuclei.

The increase in odd-even staggering away from closed shells is attributed to the greater polarizabilty of the nucleus, a decrease in the single-particle nature of the system. This is illustrated in the OES of the lead isotopes in Fig. 8.4 increasing

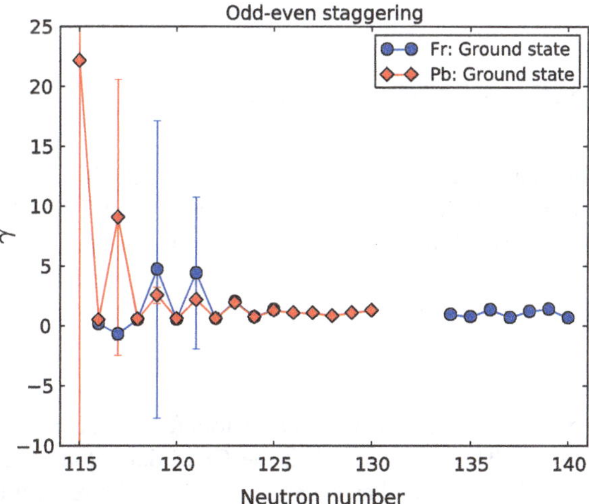

Fig. 8.4 The odd-even staggering of the francium and lead isotopes

in magnitude away from N = 126. The neutron number of the isotopes are plotted against γ, the OES parameter defined by

$$\gamma = \frac{1}{2}\frac{\delta\langle r^2\rangle_{A+1} - \delta\langle r^2\rangle_{A-1}}{\delta\langle r^2\rangle_{A} - \delta\langle r^2\rangle_{A-1}}. \tag{8.1}$$

In addition to the lead data, Fig. 8.4 presents the odd-even staggering of the francium isotopes. The heavy francium isotopes display an inversion of OES at N = 139, marking the border of reflection asymmetry. This region, from ^{221}Fr to ^{226}Fr, has a reversed odd-even staggering compared to N = 120 to 126, which is a signature for a change in the core polarization and is associated with the region of reflection asymmetry. Additional data on the neutron-rich francium isotopes ^{218,219,229,231}Fr measured during this experimental campaign has not been included. A detailed discussion on the nature of these isotopes can be found in the thesis of I. Budinčević [17] and forthcoming publications. The following discussion is limited to the neutron-deficient francium isotopes.

The francium isotope of ^{204}Fr displays a significant departure from the trend of the lead isotopes in Fig. 8.4. The reversal in the odd-even staggering at N = 118 is attributed to the larger difference in mean-square charge radii of ^{204g}Fr (with respect to ^{221}Fr) than that of ^{203}Fr. This can be seen in the charge radii of ^{203}Fr and ^{204}Fr in Fig. 8.3 where the difference in mean-square charge radius of ^{204g}Fr (compared to ^{221}Fr) equals the N = 117 lead isotope ^{199}Pb. In the case of ^{203}Fr, the difference in mean-square charge radius compared to ^{221}Fr is smaller than that of ^{198}Pb (N = 116) compared to ^{208}Pb. This causes a reversal in the OES of the francium isotopes, as shown in Fig. 8.4, a surprising result as the $\nu 3p_{3/2}$ neutron shell is depleted. The smaller mean-square charge radii for ^{203}Fr (with respect to the spherical ^{198}Fr) can be attributed to the onset of collectivity and departure from sphericity, resulting in the reversal in odd-even staggering exhibited.

8.3 Interpretation of Nuclear g-Factors

The magnetic moments of the francium isotopes were calculated in reference to the $\mu(^{210}\text{Fr})$ measurement of Gomez [18], as discussed in Sect. 7.1.7. The moments of all neutron-deficient francium isotopes measured by CRIS are shown in Fig. 8.5.

The nuclear gyromagnetic (g-) factors are calculated from the extracted magnetic moments by

$$g = \frac{\mu_I}{I}. \tag{8.2}$$

Figures 8.6 and 8.8 show the experimental g-factors for odd-A and even-A francium isotopes, respectively. These plots present the CRIS data alongside the data from Ekström [20]. The Ekström data has been re-evaluated with respect to the $\mu(^{210}\text{Fr})$ measurement of Gomez [18].

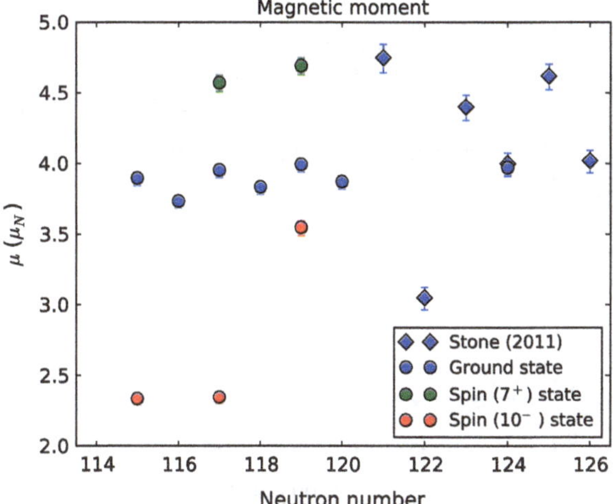

Fig. 8.5 The magnetic moments for the francium isotopes $^{202-207,211}$Fr [CRIS] and $^{208-210,212,213}$Fr [19]

In Fig. 8.6, the blue line represents the empirical g-factor the odd-A isotopes would have for the single-particle occupation of the valence proton in the $\pi 1h_{9/2}$ orbital. The g-factor for the single proton in the $\pi 1h_{9/2}$ orbital, $g_{\text{emp}}(\pi 1h_{9/2})$, was determined from the re-evaluated magnetic moment of ^{211}Fr [18, 20]. Similarly, $g_{\text{emp}}(\pi 3s_{1/2})$ was estimated from the magnetic moment of the neighbouring isotope ^{207}Tl [21].

The jump in g-factor of ^{207}Tl is thought to be associated with a significant change in the configuration mixing contribution. The coupling of the $\nu(3p^{-1}_{3/2} \otimes 3p_{1/2})_{1+}$ excitation mode (a first-order core polarization) disappears when the $\nu 3p_{3/2}$ neutron shell is filled from ^{205}Tl to ^{207}Tl. In addition, there are second-order polarization effects that are present in the neighbouring even-A lead isotopes that may result in the same trend for thallium. There is an increase in excitation energy of the first 2^+ excited state from 0.9. MeV (for N < 126) to 4.1 MeV (for N = 126) in the lead isotopes. This results in the coupling of the $(2^+ \otimes \pi d^{-1}_{3/2})_{1/2^+}$ excited state to $(0^+ \otimes \pi s^{-1}_{1/2})_{1/2^+}$ that is smaller for the N = 126 isotones. This decreased second-order core polarization effect associated with the decreased coupling of the single-hole components to the first excited 2^+ states results in an increase in magnetic moment, and thus g-factor of ^{207}Tl compared to the lighter thallium isotopes. The second-order core polarization is stronger away from N = 126 and, along with meson-exchange currents, acts to lower the magnetic moments and the resulting g-factors for N < 126.

With the depletion of the $\nu 3p_{1/2}$, $\nu 2f_{5/2}$ and $\nu 3p_{3/2}$ shells from ^{213}Fr (N = 126), there is a gradual trend away from the blue $g_{\text{emp}}(\pi 1h_{9/2})$ line. This suggests a departure from the single-particle nature of the nuclei, towards a more collective behaviour. From N = 126 to 116, every isotope has a g-factor consistent with the proton occupying the $\pi 1h_{9/2}$ orbital. This indicates that the $9/2^-$ state is still the ground state,

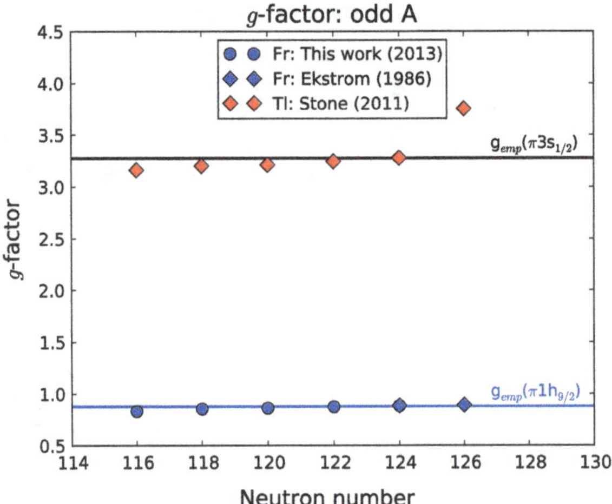

Fig. 8.6 The g-factors for the odd-A francium isotopes $^{203-213}$Fr (Z = 87) [CRIS] [18, 20] and $^{197-207}$Tl (Z = 81) isotopes [19]. The g-factors for the $\pi 3s_{1/2}$ and $\pi 1h_{9/2}$ proton orbitals have been calculated empirically. See text for details

and the $(\pi 3s^{-1}_{1/2})_{1/2^+}$ proton intruder state has not yet become the ground state. This lowering in energy of the $\pi 3s_{1/2}$ state to become the ground state would be apparent in the sudden increase in g-factor of the state, as illustrated by the black $g_{\mathrm{emp}}(\pi 3s_{1/2})$ line.

Figure 8.6 highlights the robustness of the Z = 82 and N = 126 shell closure with a shell model description valid over a range of isotopes. It also highlights the relative insensitivity of the g-factor to second-order core polarization, of which the electric quadrupole moment, Q_s is a much better probe. Figure 8.7 illustrates that the g-factor is however sensitive to bulk nuclear effects. The departure from the $g_{\mathrm{emp}}(1\pi 1h_{9/2})$ line (re-evaluated from the magnetic moment of ^{211}Fr as before [18, 20]) shows the sensitivity of the g-factor to second-order core polarization in the odd-A thallium, bismuth and francium isotopes. There is a rapidly accelerating onset in the francium isotopes as the neutron orbitals are depleted. Instead of the linear trend of the thallium and bismuth isotopes, the trend of francium has a parabolic dependence. It is suggested that the four (six) additional protons in francium, compared to bismuth (thallium) isotopes, is enough to significantly weaken the shell closure. This causes a systematic reduction of the g-factors in francium compared to the other spin $9/2^-$ isotopes presented in Fig. 8.7.

Further measurements towards the limit of stability are needed to fully understand the prediction of inversion of the $\pi 3s_{1/2}$ intruder orbital with the $\pi 1h_{9/2}$ ground state. A re-measurement of ^{203}Fr could determine the presence of the spin $1/2^+$ isomer, with a half-life of 43(4) ms [22], which was not observed during this experiment. Successful measurement of ^{202}Fr was performed during this experiment, with a yield of 100 atoms per second. By pushing the limits of laser spectroscopy, further measurements of ^{201}Fr (with a yield of 1 atom per second) and ^{200}Fr (less than 1 atom

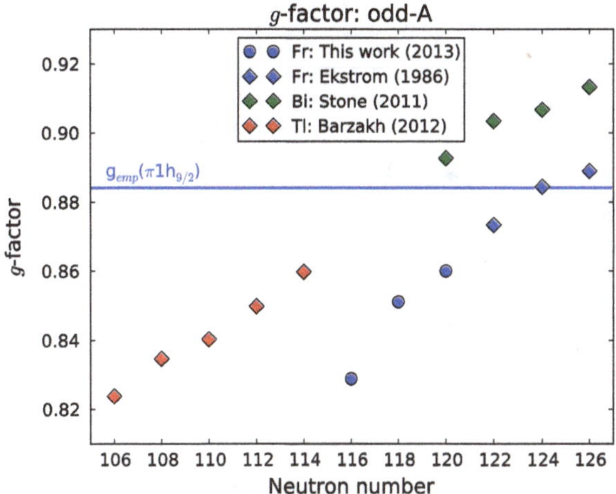

Fig. 8.7 The g-factors for the odd-A spin 9/2$^-$ isotopes: Tl (Z = 81) [23], Bi (Z = 83) [19] and Fr (Z = 87) [CRIS] [18, 20]. The g-factor for the $\pi\,1h_{9/2}$ proton orbital has been calculated empirically. See text for details

per second) are thought to be possible. The ground state (9/2$^-$) of ^{201}Fr has a half life of 53 ms and its isomer (1/2$^+$) a half life of 19 ms. By increasing the sensitivity of the CRIS technique (as detailed in Chap. 9), the presence of the 1/2$^+$ isomers in ^{201,203}Fr can be confirmed. A positive identification will lead to nuclear structure measurements that will determine (along with the verification of nuclear spin) the magnetic moments which are sensitive to the single particle structure and thus to the $(\pi 3s_{1/2})_{1/2^+}$ proton intruder nature of these states. The g-factors of ^{201m,203m}Fr will be much larger than those corresponding to the 9/2$^-$ spin states of the other francium isotopes, and will lie along the black line (denoting $g_{\rm emp}(\pi\,3s_{1/2})$) at the top of the plot.

The g-factors for the even-A francium isotopes are presented in Fig. 8.8. With the coupling of the single valence proton in the $\pi\,1h_{9/2}$ orbital with a valence neutron, a large shell model space is available. The empirically calculated g-factors for the coupling of the $\pi\,1h_{9/2}$ proton with the valence neutrons are denoted by the coloured lines. These g-factors were calculated from the additivity relation

$$g = \frac{1}{2}\Big[g_p + g_n + (g_p - g_n)\frac{j_p(j_p+1) - j_n(j_n+1)}{I(I+1)}\Big], \tag{8.3}$$

as outlined by Ekström [20]. The empirical g-factors of the odd valence neutrons were calculated from the magnetic moments of neighbouring nuclei: ^{201}Po for the blue $g_{\rm emp}(\pi\,1h_{9/2} \otimes \nu 3p_{3/2})$ and red $g_{\rm emp}(\pi\,1h_{9/2} \otimes \nu 1i_{13/2})$ line [24]; ^{213}Ra for the black $g_{\rm emp}(\pi\,1h_{9/2} \otimes \nu 3p_{1/2})$ line; and ^{211}Ra for the green $g_{\rm emp}(\pi\,1h_{9/2} \otimes \nu 2f_{5/2})$ line [25]. The empirical g-factors for the valence proton in the $\pi\,1h_{9/2}$ orbital were

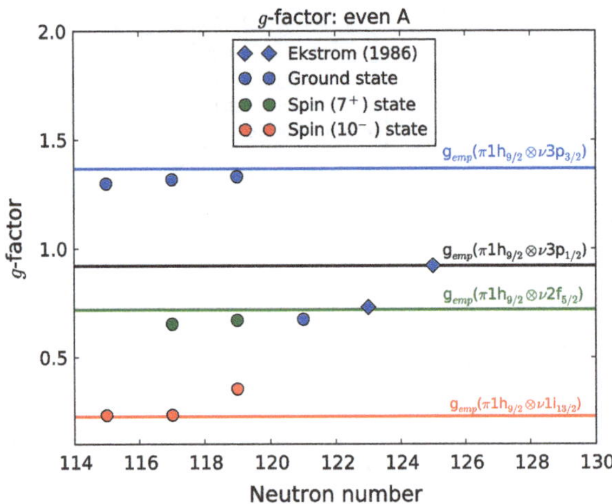

Fig. 8.8 The g-factors for the even-A francium isotopes $^{202-212}$Fr [CRIS] [18, 20]. The g-factor for the coupling of the proton and neutron orbitals have been calculated empirically. See text for details

calculated from the magnetic moment of the closest odd-A francium isotope (^{203}Fr and ^{211}Fr respectively) from the CRIS data.

The ground state of ^{202,204,206}Fr display similar g-factors, with the valence proton and neutron coupling to give a spin 3^+ state. The tentative configuration in literature of $(\pi 1h_{9/2} \otimes \nu 3f_{5/2})_{3^+}$ for ^{202g}Fr is based on the configuration of the (3^+) state in ^{194}Bi [26]. Similarly, the assignment of the same configuration for ^{204g}Fr and ^{206g}Fr is based on the systematics of neighbouring nuclei ^{196,198}Bi (including the corresponding decays within the ^{204}Fr-^{200}At-^{196}Bi and ^{206}Fr-^{202}At-^{198}Bi decay chains). However, the initial assignment of ^{194g}Bi was declared to be either $(\pi 1h_{9/2} \otimes \nu 3f_{5/2})_{3^+}$ or $(\pi 1h_{9/2} \otimes \nu 3p_{3/2})_{3^+}$ [27]. From the g-factors of the ground state and isomers of ^{202,204,206}Fr, it is suggested that the configuration of these states is instead $(\pi 1h_{9/2} \otimes \nu 3p_{3/2})_{3^+}$.

The first isomeric states of ^{204,206}Fr (7^+) have a valence neutron that occupies the $\nu 2f_{5/2}$ state. This coupling of the proton-particle neutron-hole results in a $(\pi 1h_{9/2} \otimes \nu 2f_{5/2})_{7^+}$ configuration [28]. For ^{202m2}Fr, ^{204m2}Fr and ^{206m2}Fr, the particle proton-neutron hole coupling result in a tentative $(\pi 1h_{9/2} \otimes \nu 1i_{13/2})_{10^-}$ configuration assignment for each isomer [29]. However, while the agreement of the g-factors of the spin 10^- state in ^{202,204}Fr point to a $\nu 1i_{13/2}$ occupancy, they are in disagreement with the g-factor of the 10^- state in ^{206}Fr. The charge radius of ^{206m2}Fr indicates a highly deformed, collective state, where the single-particle description of the nucleus is no longer valid. This is consistent with the g-factor of this state: it is no longer obeying the shell model description. This leads to the conclusion, that while a $(\pi 1h_{9/2} \otimes \nu 1i_{13/2})_{10^-}$ configuration for ^{206m2}Fr is suggested, the charge radii and magnetic moment point to a drastic change in the structure of the 10^- state.

Table 8.1 Tentative proton and neutron configurations for $^{202-213}$Fr

Isotope	Spin	Configuration		References
		π	ν	
202g	(3^+)	$1h_{9/2}$	$3p_{3/2}$	[CRIS]
202m	(7^+)	$1h_{9/2}$	$1i_{13/2}$	[26]
203	$(9/2^-)$	$1h_{9/2}$		[CRIS]
204g	(3^+)	$1h_{9/2}$	$3p_{3/2}$	[CRIS]
204m1	(7^+)	$1h_{9/2}$	$2f_{5/2}$	[29]
204m2	(10^-)	$1h_{9/2}$	$1i_{13/2}$	[29]
205	$(9/2^-)$	$1h_{9/2}$		[30]
206g	(3^+)	$1h_{9/2}$	$3p_{3/2}$	[CRIS]
206m1	(7^+)	$1h_{9/2}$	$2f_{5/2}$	[28]
206m2	(10^-)	$1h_{9/2}$	$1i_{13/2}$	[28]
207	$9/2^-$	$1h_{9/2}$		[20]
208	7^+	$1h_{9/2}$	$2f_{5/2}$	[31]
209	$9/2^-$	$1h_{9/2}$		[20]
210	6^+	$1h_{9/2}$	$2f_{5/2}$	[31]
211	$9/2^-$	$1h_{9/2}$		[20]
212	5^+	$1h_{9/2}$	$3p_{1/2}$	[32]
213	$9/2^-$	$1h_{9/2}$		[20]

For completeness, the configurations of the even-A francium isotopes ^{208,210,212}Fr are presented. The coupling of the valence proton and neutron in the $\pi 1h_{9/2}$ and $\nu 2f_{5/2}$ orbital in ^{208}Fr and ^{210}Fr leads to a $(\pi 1h_{9/2} \otimes \nu 2f_{5/2})_{7^+}$ and $(\pi 1h_{9/2} \otimes \nu 2f_{5/2})_{6^+}$ configuration respectively [31]. With the $\nu 2f_{5/2}$ neutron orbital now full, the valence neutron in ^{212}Fr occupies the $\nu 3p_{1/2}$ orbital, resulting in a $(\pi 1h_{9/2} \otimes \nu 3p_{1/2})_{5^+}$ configuration [32].

The agreement of the experimental and empirical g-factors, as shown in Figs. 8.6, 8.7 and 8.8, illustrates the suitability of the shell model description of the neutron-deficient francium isotopes. These isotopes display a single-particle nature where the additivity relation is still reliable. The tentative proton and neutron configurations for the francium isotopes are displayed in Figs. 8.6 and 8.8 and are presented in Table 8.1.

8.4 Alpha-Decay Systematics

The occurrence of shape coexistence in the neutron-deficient isotopes in the region of $Z = 82$ has generated a lot of interest, both experimentally and theoretically [33, 34]. The even-even nuclei have benefitted from alpha- and beta-decay studies that have probed this phenomenon induced by proton excitations across the shell gap [35, 36]. In the odd-A and odd-odd nuclei, these excitations may cause isomerism.

8.4.1 The Odd-A Francium Isotopes

The systematic trends in the neutron-deficient bismuth (Z = 83) and astatine (Z = 85) isotopes have been well characterized, with the ground state isotopes occupying the $(\pi 1h_{9/2})_{9/2^-}$ state with a spin of $9/2^-$. The $(\pi 3s^{-1}_{1/2})_{1/2^+}$ proton intruder state in the neutron-deficient bismuth isotopes decreases in energy, becoming the ground state at ^{185}Bi (N = 102) [37]. A similar mechanism has been observed in the astatine isotopes, where the $1/2^+$ isomer was first discovered with the alpha decay of ^{197}At [38]. This low-lying intruder state is the ground state in ^{195}At (N = 110). The presence of an $1/2^+$ isomer was first observed in the francium isotopes in ^{201}Fr, tentatively in ^{203}Fr [39] and recently in ^{205}Fr [40]. It has been suggested that this $(\pi 3s^{-1}_{1/2})_{1/2^+}$ proton intruder state becomes the ground state in ^{199}Fr (N = 112).

The excitation energy of the $1/2^+$ state in ^{201}Fr is inconsistent in literature, and has been estimated to be 146 keV [39] and 129 keV [42], with no quoted errors. With no observation of the alpha decay of the corresponding state in ^{199}At, the excitation energy of the isomer in ^{203}Fr could not be determined [39]. Thus, a renewed effort in recent years has been made to characterise this region with alpha-decay studies [22, 40, 41, 43]. The current state-of-knowledge of the systematics of the $(\pi 3s^{-1}_{1/2})_{1/2^+}$ proton intruder state is summarized in Fig. 8.9.

Spectroscopy of the neutron-deficient francium isotopes produced in fusion evaporation reactions at RITU (Jyväskylä, Finland) [22, 40, 41] and SHIP (GSI, Germany) [43] have recently been performed. Such studies are investigating proton drip-line nuclei, where ^{201}Fr is the first proton-unbound francium isotope (in one-proton separation energy) [44]. Following the discovery of the $1/2^+$ state in ^{199}At ($t_{1/2}$ = 0.31(8) s), the excitation energy of the proton intruder state in ^{203}Fr has been determined to be ∼360 keV [22] with a half-life of 43(4) ms. Furthermore, a long-lived ($t_{1/2}$ = 1.15(4) ms) isomeric state in ^{205}Fr has been observed. With a spin-parity of $1/2^+$, the identity has been suggested as the proton intruder orbital, but the excitation energy of this state is tentative [40].

^{199}Fr	^{201}Fr	^{203}Fr	^{205}Fr
<300 keV, 1.6 ms $(13/2^+)$	$(7/2^-)$	∼360 keV, 43(4)ms $(1/2^+)$	∼609 keV, 1.15(4) ms $(1/2^+)$
56 keV, 7ms $(7/2^-)$	145 keV, $19(^{+19}_{-6})$ ms $(1/2^+)$	162 keV $(7/2^-)$	209 keV $(7/2^-)$
0 keV, 5 ms $(1/2^+)$	53(4) ms $(9/2^-)$	0.53(2) s $(9/2^-)$	3.97(4) s $(9/2^-)$

Fig. 8.9 Schematic illustration of the $(\pi 3s^{-1}_{1/2})_{1/2^+}$ proton intruder state in the neutron-deficient odd-A francium isotopes [22, 39–41]

Recent alpha-decay spectroscopy of ^{199}Fr ($t_{1/2}=5$ ms) has suggested the inversion of the ground state of the heavier francium isotopes (9/2$^-$) with that of the proton intruder state, proposing a ground-state spin of 1/2$^+$. In the energy level scheme of this isotope, the 7/2$^-$ state is now 56 keV above the ground state, with a higher-lying (<300 keV) 13/2$^+$ state also observed [41]. However, due to the low statistics it was difficult to determine the origin of the three alpha particles observed. Despite this, the very existence of the three alpha-decay peaks is indicative of a change in the low-energy structure of ^{199}Fr, compared to the $N>112$ isotopes. This supports the previous prediction of an onset of ground-state deformation at $N=112$.

However, an additional alternative interpretation is presented in Ref. [43]. The energy of the alpha particles emitted from the decay of the 1/2$^+$ and 7/2$^-$ states (7,664(11) and 7,676(6) keV, respectively) have a large statistical uncertainty. It is possible that the 7/2$^-$ state, and not the 1/2$^+$ state, is the ground state of ^{199}Fr, with a half-life of 6(1) ms. In this scheme, the ^{195}At ground state (spin 1/2$^+$, $t_{1/2}=328(20)$ ms) would be populated by the E3 internal transition from the 7/2$^-$ state, instead of from the alpha decay of the 1/2$^+$ state in ^{199}Fr. Consequently, the presence or absence of the proton intruder state in ^{199}Fr is still under discussion. On account of the low-spin 1/2$^+$ state being weakly populated in heavy ion induced fusion-evaporation reactions, knowledge of this region would benefit hugely from further measurements at radioactive beam facilities that favour the creation of low-spin states, at ISOL facilities such as ISOLDE.

Spectroscopic studies on the very neutron-deficient francium isotopes have observed the lightest francium isotope to date, ^{197}Fr. This isotope was identified with a 0.6(3) ms half-life and a spin-parity of 7/2$^-$, based on the systematics of the 7/2$^-$ state in ^{194}At. The unobserved 1/2$^+$ state in ^{197}Fr could be due to the aforementioned tendency to populate higher-spin states in comparison to those of lower-spin in fusion-evaporation reactions [43].

From the charge-radii measurements, as discussed in Sect. 8.1, there appears to be an onset of deformation in the francium isotopes as early as ^{203}Fr ($N=116$) which is shown in Fig. 8.3. The g-factors of the odd-A francium isotopes lie further away from the $g_{\mathrm{emp}}(\pi 1h_{9/2})$ single-particle configuration which is consistent with an onset of collectivity. This however is associated with the bulk properties of the nucleus, and does not represent the change in the nature of the ground state of the nucleus which may remain essentially spherical in these isotopes. From the recent alpha-decay studies, static ground-state deformation has been proposed to occur at ^{199}Fr. Given the sensitivity of the hyperfine structure of an isotope (and isomer) to the nuclear configuration of the state, laser spectroscopy measurements on francium isotopes ^{201}Fr, ^{200}Fr and ^{199}Fr would contribute significantly to the knowledge of the underlying characteristics of these neutron-deficient isotopes. High resolution measurements would allow the upper state splitting $B_{P_{3/2}}$ of the 8p $^2P_{3/2}$ electronic orbital to be measured. This would provide the electric quadrupole moment Q_s of the nucleus (see Sect. 2.1.3), and the static deformation of the nucleus to be determined.

The experimental challenge facing such measurements is the sensitivity of the technique. The francium isotope ^{202}Fr was produced during the last experiment at a rate of less than 10^2 atoms per second, and it is estimated the yield of ^{201}Fr to be of the order of 1 atom per second. With the standard collinear laser spectroscopy technique limited to at least 10^3 atoms per second, the CRIS technique offers the only opportunity to measure the hyperfine structure of the ground and isomeric states of these rare isotopes, and provide the necessary nuclear structure information to understand their configuration.

8.4.2 The Even-A Francium Isotopes

The current state-of-knowledge of the low-lying (less than 1 MeV excitation energy) structure of the isomers in the even-A francium isotopes is summarized in Fig. 8.10. There are three low-lying (less than 1 MeV) isomers in ^{206}Fr that have been identified through their Fr-At-Bi alpha-decay chain [45]. The spin-parity assignments of these three states are based on the decay patterns and systematics of the β^+/EC-decaying states in ^{198}Bi. These transitions are thought to decay to the same-spin parity states in ^{202}At, as suggested by their small hindrance factors. In addition there is a E3 internal transition of 531 keV from the (10^-) to the (7^+) state.

A similar pattern is observed in ^{204}Fr, with the three isomeric states identified with the corresponding spin-parity. The radioactive decay of ^{204}Fr and its low-lying isomers is shown in Fig. 8.11. Figure 7.17 illustrates the different energies the emitted alpha particles have when the laser is tuned onto resonance with each of the three isomeric states. The alpha-particle energy spectrum in Fig. 7.21 illustrates the presence of the 6,969 keV alpha particle emitted from the decay of ^{204m1}Fr. This suggests the decay of the (10^-) state in ^{204m2}Fr to the (7^+) state. While the E3 internal transitions from the (10^-) to the (7^+) state in ^{200}At and ^{196}Bi have been observed, the complementary transition in ^{204m2}Fr was not observed during the initial discovery of the isomeric states in ^{204}Fr [45]. Despite the excitation energy being predicted from

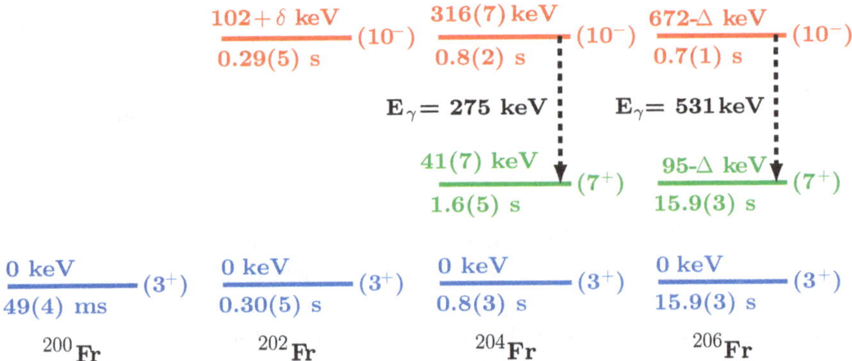

Fig. 8.10 Schematic illustration of the (3^+), (7^+) and (10^-) isomeric states in even-A francium isotopes [40, 42, 45, 46]

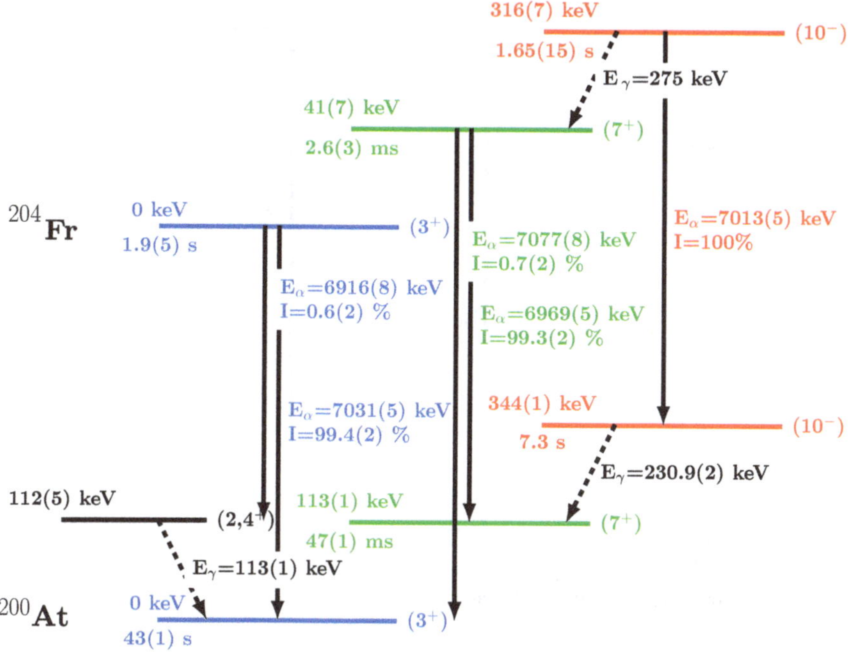

Fig. 8.11 The radioactive decay of ^{204}Fr and its isomers [40, 45, 46]

the alpha-decay energies, it was not observed until 20 years later with electron-alpha decay tagging [40].

Based on the (3^+) and (10^-) states in ^{194}Bi, the spin-parity of the two observed states in ^{202}Fr are given the same assignments. Despite only one alpha-particle energy having been observed (7,237 keV, $t_{1/2}=0.34$ s), both the (3^+) and (10^-) states in ^{198}At were populated. This led to the assumption that there is a doublet structure in ^{202}Fr that was later confirmed [39]. The energy of the emitted alpha particles and half-life of the states were remeasured to be $E_\alpha(3^+)=7{,}241(8)$ keV, $t_{1/2}=0.30(5)$ s and $E_\alpha(10^-)=7{,}235(8)$ keV, $t_{1/2}=0.29(5)$ s respectively. The radioactive decay of ^{202}Fr and its low-lying isomer is shown in Fig. 8.12.

The DSS was used to identify the ground state and isomeric state of ^{202}Fr after an initial hyperfine structure scan was performed. The laser was tuned onto resonance with two of the hyperfine structure peaks and a radioactive decay measurement was performed for 15 min. Since only a short time was spent on resonance (for identification purposes), low statistics from the CRIS experiment meant that the laser assisted nuclear decay spectroscopy of ^{202}Fr could not provide a more accurate determination of the energy and half-lives of these two states. However, Fig. 7.27 clearly shows a difference in energy of the daughter alpha particles emitted from ^{198g}At and ^{198m}At. With a difference in the mean energy of the alpha particles associated with ^{202g}Fr and ^{202m}Fr of 8 keV, the spectrum suggests that the two states

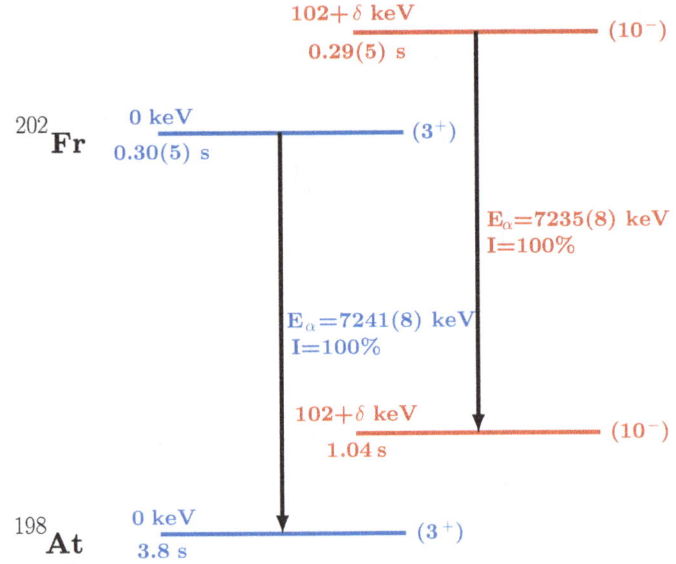

Fig. 8.12 The radioactive decay of ^{202}Fr and its isomer [42, 46]

of ^{202}Fr also have slightly different energies. A repeat measurement of laser assisted nuclear decay spectroscopy with higher statistics would allow the CRIS experiment to contribute to the characteristics of the isomeric states of ^{202}Fr.

The low-lying structure of ^{200}Fr has been a topic of recent discussion, with the alpha-particle energy and half-life of the ground state having inconsistent values in literature. Fusion evaporation reactions at GARIS (RIKEN, Japan) and RITU (Jyväskylä, Finland) observed the alpha decay of ^{200}Fr, but with very different values: $E_\alpha = 7{,}500(30)$ keV, $t_{1/2} = 570^{+270}_{-140}$ ms [47] and $E_\alpha = 7{,}468(9)$ keV, $t_{1/2} = 19^{+13}_{-6}$ ms [48] respectively. With only six observed alpha-decay chains for each experiment, the low statistics motivated a new study of the radioactive decay of ^{200}Fr at ISOLDE (CERN, Switzerland) with improved accuracy [42, 49]. Only the ground state in ^{200}Fr ($t_{1/2} = 49(4)$ ms) was observed with an alpha decay of energy $E_\alpha(3^+) = 7{,}473(12)$ keV, consistent within the experimental errors of the previous measurements. Although a low-lying isomer cannot be ruled out, the singularly observed alpha branch into the ground state of ^{196}At makes this hypothesis unlikely. Performing collinear resonance ionization spectroscopy of ^{200}Fr would confirm the existence of an isomer, if one was present.

The energy of the (10^-) to (7^+) E3 transitions in bismuth, astatine and francium are compared to the energy difference of the $13/2^+$ and $5/2^-$ states in odd-A lead isotopes (arising from a valence neutron in the $\nu 1i_{13/2}$ and $\nu 3f_{5/2}$ orbitals) in Fig. 8.13. As discussed in Sect. 8.3 (see Table 8.1), the (10^-) and (7^+) states result from the $(\pi 1h_{9/2} \otimes \nu 1i_{13/2})_{10^-}$ and $(\pi 1h_{9/2} \otimes \nu 2f_{5/2})_{7^+}$ configurations respectively. The remarkable similarity in the energy of the $13/2^+$ to $5/2^-$ transition in lead to the

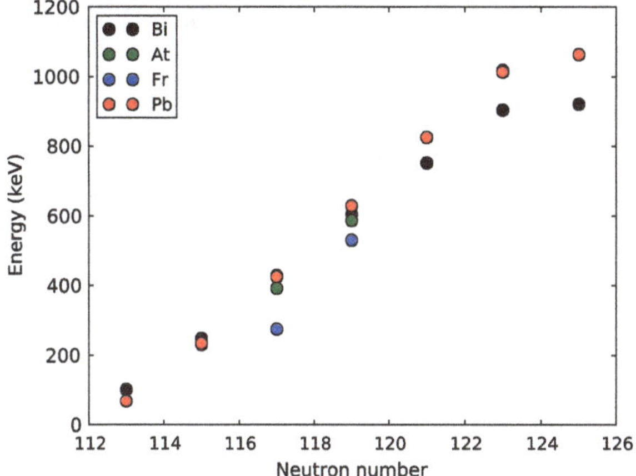

Fig. 8.13 The energy of the (10^-) to (7^+) E3 transitions in Bi, At and Fr compared to the energy difference of the 13/2$^+$ and 5/2$^-$ states in odd-A lead isotopes [45, 46]

(10^-) to (7^+) transition in bismuth suggests that the valence proton in bismuth acts as a spectator particle, and that the coupling of the proton and neutron remains constant as the neutron number increases [45]. The addition of two protons (in the case of astatine) and four protons (for francium) does not change the transition energy of these states. Since the identification of the E3 internal transition in ^{204}Fr [40], this data point can be included in the plot. This is illustrated in Fig. 8.13, where the 275 keV transition in ^{204}Fr shows a deviation from the trend.

The magnetic moments of the (10^-) states in ^{204,206}Fr (Fig. 8.8) suggest that ^{204m2}Fr has a pure $(\pi 1h_{9/2} \otimes \nu 1i_{13/2})_{10^-}$ configuration. In contrast, the ^{206m2}Fr g-factor diverges from the $g_{\mathrm{emp}}(\pi 1h_{9/2} \otimes \nu 1i_{13/2})$ line, hereby deviating from the single-particle configuration and a change in the nature of the state is suggested. However, from the radioactive decay systematics shown in Fig. 8.13, it is ^{204}Fr that shows a departure from the systematic trend of bismuth and astatine, indicating the valence proton no longer acts as a spectator particle.

8.4.3 Q_α Systematics

Despite the general trend for the Q-value of alpha decay to increase with mass number, the local systematic trend of decreasing Q_α for the francium isotopes from N = 112 to 119 is shown in Fig. 8.14. It illustrates the decrease of Q_α with increasing neutron number as the shell closure at N = 126 is approached. After this shell closure, the Q-value rapidly increases, a fingerprint of increased stability of the magic nuclei. Figure 8.14 presents the close agreement of all the neutron-deficient francium isotopes in this region with this trend. Information for ^{206}Fr to ^{199}Fr is presented [46].

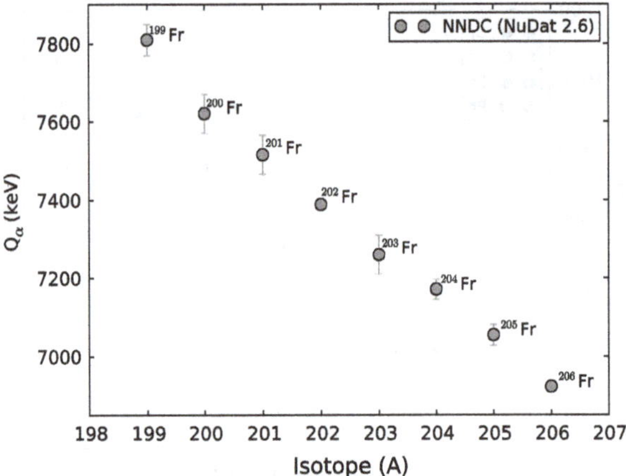

Fig. 8.14 The systematic decrease in Q_α as the francium isotopes approach the N = 126 shell closure. The literature values for Q_α are taken from Ref. [46]

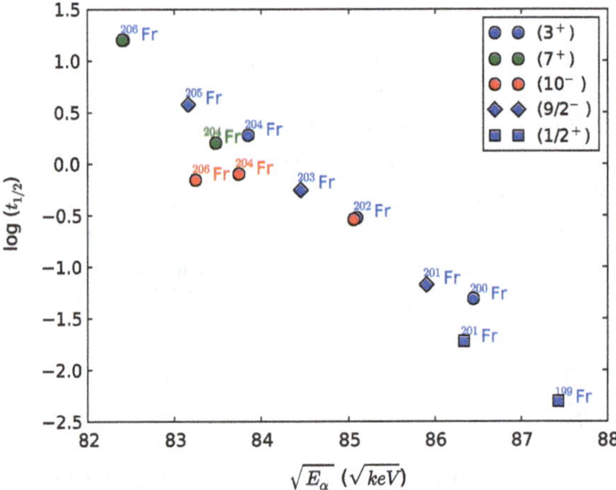

Fig. 8.15 A Geiger-Nuttall plot for neutron-deficient francium isotopes. The literature values for E_α and $t_{1/2}$ are taken from Refs. [41, 42, 46]

The Geiger-Nutall law of alpha decay [50] relates the Q-value with the half-life of the decaying state,

$$\log(t_{1/2}) = a + \frac{b}{\sqrt{Q_\alpha}}, \tag{8.4}$$

where the constants a and b have a Z dependence. A 1 MeV change in the alpha-decay energy corresponds to a change in the half-life of the state of the order of 10^5.

Figure 8.15 shows the Geiger-Nuttall plot for the neutron-deficient francium isotopes $^{199-206}$Fr. The agreement of the francium isotopes with the Geiger-Nuttall law is good, with the exception of ^{206m2}Fr which again points to the unusual behaviour of this isomer. The anomalies in the alpha-decay energies of the neutron-deficient polonium (Z = 84) isotopes have been suggested to be due to the extra binding energy arising from deformation. From ^{191}Po (N = 107) to ^{189}Po (N = 105), theory predicts a shape change in the nucleus from oblate to prolate. The stronger binding energy of the prolate state reduces the Q_α transition energy of the alpha decay and would explain the deviation from the Geiger-Nuttall law in the polonium isotopes. By the same argument, it is suggested that the departure of ^{206m2}Fr, as illustrated in Fig. 8.15, is a result of the significant deformation of the (10^-) state. This is in good agreement with the conclusions drawn from the mean-square charge radius measurement of ^{206m2}Fr, as discussed in Sect. 8.1. In addition, there is a scatter in the values of the isomeric states of ^{204}Fr that is not present for other isotopes with low-lying states, for example ^{202g,m}Fr and ^{206g,m1}Fr. This again suggests an onset of deformation in the (7^+) and (10^-) states.

References

1. Myers WD, Schmidt KH (1983) Nucl Phys A 410:61
2. Berdichevsky D, Tondeur F, Phys Z (1985) A 322:141
3. Goddard PM et al (2013) Phys Rev Lett 110:032503
4. Ulm G et al (1986) Z Phys A 325:247
5. Anselment M et al (1986) Nucl Phys A 451:471
6. De Witte H et al (2007) Phys Rev Lett 98:112502
7. Fricke G, Heilig K (2005) Nuclear charge radii. Springer, Berlin
8. Cocolios TE et al (2011) Phys Rev Lett 106:052503
9. Borchers W et al (1987) Hyperfine Interact 34:25
10. Dzuba VA et al (2005) Phys Rev A 72:022503
11. Wansbeek LW et al (2012) Phys Rev C 86:015503
12. Nazarewicz W et al (1984) Nucl Phys A 429:269
13. Voss A et al (2013) Application of lasers and storage devices in atomic nuclei research. Poznan, Poland
14. Tomlinson WJ, Stroke HH (1962) Phys Rev Lett 8:436
15. Talmi I (1984) Nucl Phys A 423:189
16. Zawischa D (1985) Phys Lett B 155:309
17. Budinčević I (2014) Ph.D. thesis, IKS, KU Leuven
18. Gomez E et al (2008) Phys Rev Lett 100:172502
19. Stone N (2011) Table of nuclear magnetic dipole and electric quadrupole moments. Nuclear data services, International Atomic Energy Agency, Vienna
20. Ekström C et al (1986) Phys Scripta 34:624
21. Neugart R et al (1985) Phys Rev Lett 55:1559
22. Jakobsson U et al (2013) Phys Rev C 87:054320
23. Barzakh AE et al (2012) Phys Rev C 86:014311
24. Wouters J et al (1991) J Phys G Nucl Part Phys 17:1673
25. Ahmad S et al (1983) Phys Lett B 133:47
26. Zhu S, Kondev F (2008) Nucl Data Sheets 109:699
27. Duppen PV et al (1991) Nucl Phys A 529:268

28. Kondev F (2008) Nucl Data Sheets 109:1527
29. Chiara C, Kondev F (2010) Nucl Data Sheets 111:141
30. Kondev F (2004) Nucl Data Sheets 101:521
31. Martin M (2007) Nucl Data Sheets 108:1583
32. Browne E (2005) Nucl Data Sheets 104:427
33. Andreyev AN (2000) Nature 405:430
34. Van Duppen P, Huyse M (2000) Hyperfine Interact. 129:149
35. Wood J et al (1992) Phys Rep 215:101
36. Heyde K, Wood JL (2011) Rev Mod Phys 83:1467
37. Andreyev AN et al (2004) Phys Rev C 69:054308
38. Coenen E et al (1986) Z Phys A 324:485
39. Uusitalo J et al (2005) Phys Rev C 71:024306
40. Jakobsson U et al (2012) Phys Rev C 85:014309
41. Uusitalo J et al (2013) Phys Rev C 87:064304
42. De Witte H et al (2005) Eur Phys J A 23:243
43. Kalaninová Z et al (2013) Phys Rev C 87:044335
44. Uusitalo J et al (2011) AIP Conf Proc 1409:25
45. Huyse M et al (1992) Phys Rev C 46:1209
46. http://www.nndc.bnl.gov/nudat2/
47. Morita K et al (1995) Z Phys A 352:7
48. Enqvist T et al (1996) Z Phys A 354:1
49. Keupers M (2010) Ph.D. thesis, IKS, KU Leuven
50. Geiger H, Nuttall J (1911) Philos Mag 22:613

Chapter 9
Conclusion

This thesis presents the spectroscopic studies of the neutron-deficient francium isotopes performed by the CRIS experiment at ISOLDE, CERN. The hyperfine structure of $^{202-207,211,220}$Fr with reference to ^{221}Fr was measured with collinear resonance ionisation spectroscopy and the isotope shifts, mean-square charge radii and magnetic moments extracted. The selectivity of the resonance ionization process allowed laser assisted nuclear decay spectroscopy to be performed on ^{202,204,218m}Fr.

The resonant atomic transition of 7s $^2S_{1/2} \rightarrow$ 8p $^2P_{3/2}$ was probed, and the hyperfine factor $A_{S_{1/2}}$ was measured. A King plot analysis of the 422.7 nm transition in francium allowed the atomic factors to be extracted for the first time. The field and mass factors were determined to be $F_{422} = -20.668(214)$ GHz/fm^2 and $M_{422} = +752(327)$ GHz.amu, respectively.

The novel technique of laser assisted nuclear decay spectroscopy was performed on the isotopes ^{202,204,218m}Fr. In addition to performing alpha-decay spectroscopy on pure ground and isomeric states, the decay spectroscopy station was utilized to identify the peaks in the hyperfine spectra of ^{202,204}Fr. Alpha-tagging the hyperfine structure scan of ^{204}Fr allowed the accurate determination of the nuclear observables of the three low-lying isomeric states. A half life of the isomeric state of ^{218m}Fr was determined to be 20.9(4) ms.

Analysis of mean-square charge radii suggests an onset of collectivity that occurs at $N = 116$ for ^{203}Fr. However, measurement of the spectroscopic quadrupole moment is required to determine the nature of the deformation (static or dynamic). The magnetic moments suggest that the shell-model description of the neutron-deficient francium isotopes still holds, with the exception of the (10^-) isomeric state of ^{206m2}Fr. Based on the systematics of the region, the tentative assignment of the hyperfine structure peaks in ^{206}Fr result in magnetic moments and charge radii that suggest a highly deformed state. This is supported by the usual behaviour of ^{206m2}Fr in the alpha-decay systematics. A repeat measurement of collinear resonance ionization spectroscopy of ^{206}Fr would confirm the presence of the isomeric states, while laser assisted nuclear decay spectroscopy of ^{206}Fr would unambiguously determine their identity.

K.M. Lynch, *Laser Assisted Nuclear Decay Spectroscopy*,
Springer Theses, DOI: 10.1007/978-3-319-07112-1_9

The occupation of the valence proton in the $(\pi 1h_{9/2})_{9/2^-}$ orbital has been suggested for all measured isotopes down to ^{202}Fr, indicating the $(\pi 1s^{-1}_{1/2})_{1/2^+}$ proton intruder state has not yet inverted with the ground state. Further measurements of the very neutron-deficient francium isotopes towards ^{199}Fr are required to fully determine the nature of the proton intruder state. A positive identification will lead to nuclear structure measurements that will determine (along with verification of the nuclear spin), the magnetic moment which is sensitive to the single particle structure and thus to the $(\pi 1s^{-1}_{1/2})_{1/2^+}$ proton intruder nature of the state. A laser line width of 1.5 GHz was enough to resolve the lower state splitting of the hyperfine structure and measure the $A_{S_1/2}$ factor. In the future, the inclusion of a narrow-band laser system for the resonant excitation step will be able to resolve the upper state splitting, providing the $B_{P_{3/2}}$ factor. This will allow extraction of the spectroscopic quadrupole moment and determination of the nature of the deformation.

In addition, optimization of the laser beam through the CRIS experimental beam line will limit the reflection of the 1,064 nm laser light in the vacuum chambers. This will reduce the fluctuation of the baseline of the signal from the silicon detector and improve the energy resolution for future alpha-decay spectroscopy. A new detection chamber has been designed to increase the gamma-ray detection efficiency, allowing placement of the germanium detectors in closer geometry than previously allowed.

The successful measurements performed by the CRIS experiment demonstrates the high sensitivity of the collinear resonance ionization technique. The presence of the decay spectroscopy station provides the ability to identify overlapping hyperfine structure and perform radioactive decay measurements on pure ground and isomeric state beams. These two complementary techniques will be utilized in the proposed experimental campaigns to measure copper [1], polonium [2], gallium [3] and the additional francium isotopes [4]. The initial success of the CRIS experiment signifies an exciting time for high sensitivity, high resolution laser spectroscopy experiments.

References

1. Neyens G et al (2011) CERN-INTC-2011-052 INTC-P-316, CERN, Geneva
2. Cocolios TE et al (2012) CERN-INTC-2012-025 INTC-I-145, CERN, Geneva
3. Cocolios TE et al (2013) CERN-INTC-2013-010 INTC-P-375, CERN, Geneva
4. Flanagan K et al (2014) CERN-INTC-2014-020 INTC-P-240-ADD-1, CERN, Geneva

Appendix A
Collinear Resonance Ionization Spectroscopy Scans

The following figures are example collinear resonance ionization spectroscopy scans of the francium isotopes taken during the experimental campaign see Figs. A.1– A.8. The even-A francium scans of ^{202}Fr, ^{204}Fr and ^{206}Fr present the summed fit of the ground (3^+), first isomeric (7^+) and second isomeric (10^-) states, see Figs. A.1, A.3 and A.5 respectively. The odd-A francium scans of ^{203}Fr, ^{205}Fr, ^{207}Fr, ^{211}Fr and ^{220}Fr are included for completeness, see Figs. A.2, A.4, A.6–A.8, respectively.

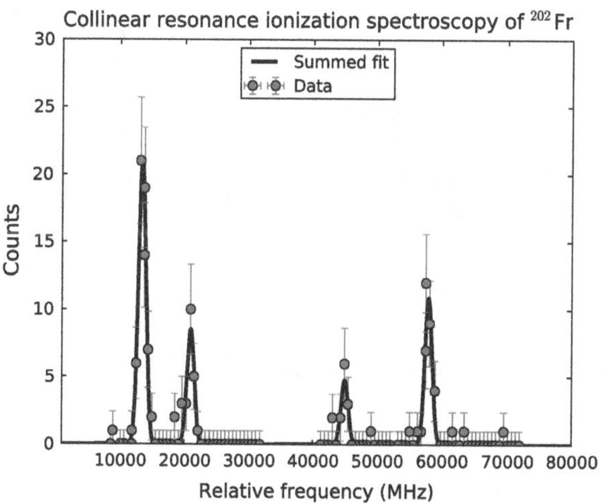

Fig. A.1 Collinear resonance ionization spectroscopy scan of ^{202}Fr

K.M. Lynch, *Laser Assisted Nuclear Decay Spectroscopy*,
Springer Theses, DOI: 10.1007/978-3-319-07112-1

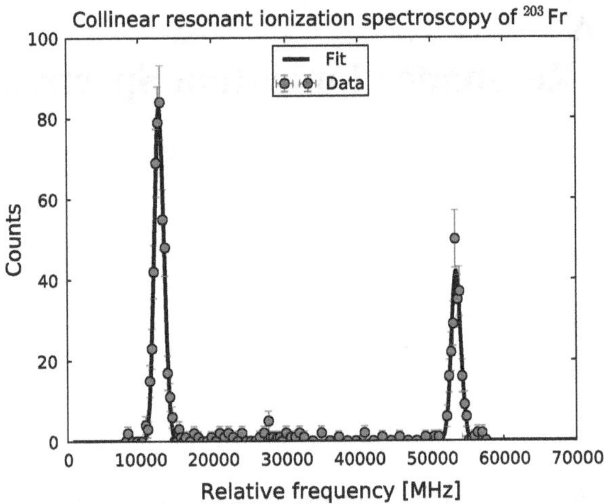

Fig. A.2 Collinear resonance ionization spectroscopy scan of ^{203}Fr

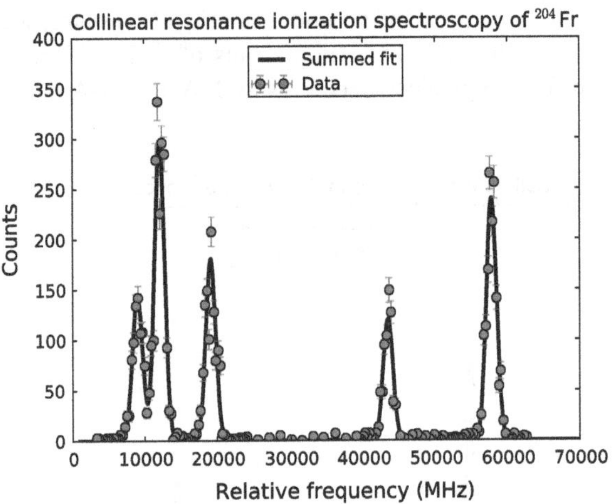

Fig. A.3 Collinear resonance ionization spectroscopy scan of ^{204}Fr

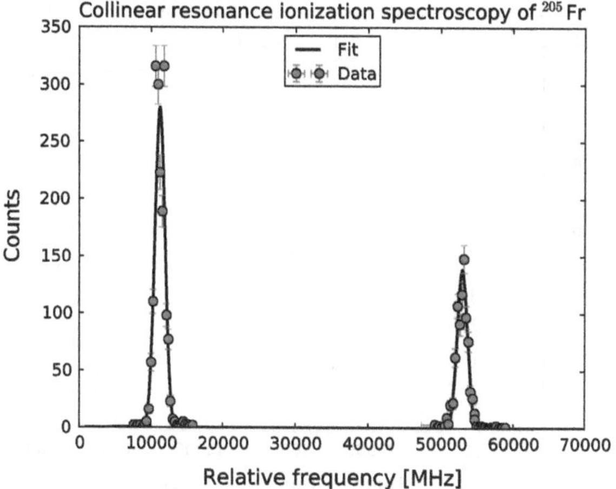

Fig. A.4 Collinear resonance ionization spectroscopy scan of ^{205}Fr

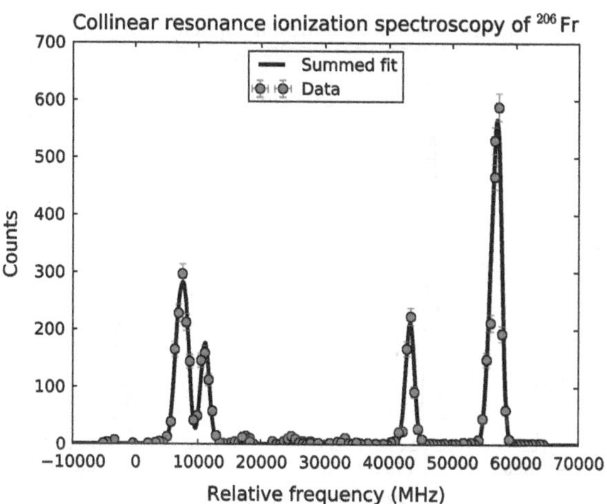

Fig. A.5 Collinear resonance ionization spectroscopy scan of ^{206}Fr

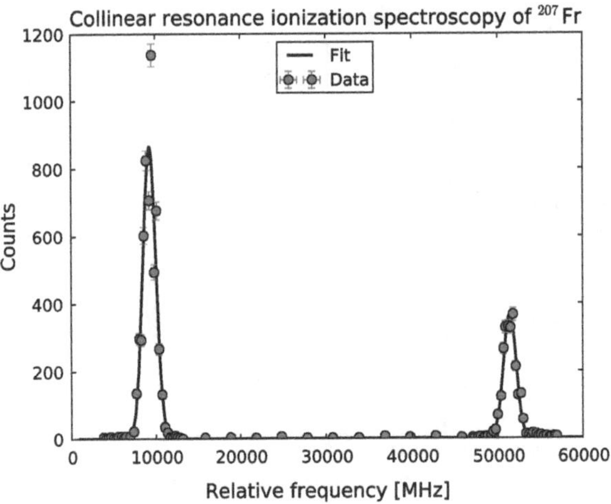

Fig. A.6 Collinear resonance ionization spectroscopy scan of ^{207}Fr

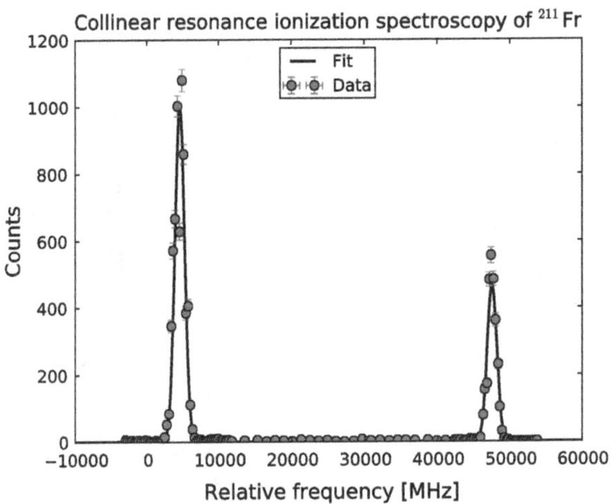

Fig. A.7 Collinear resonance ionization spectroscopy scan of ^{211}Fr

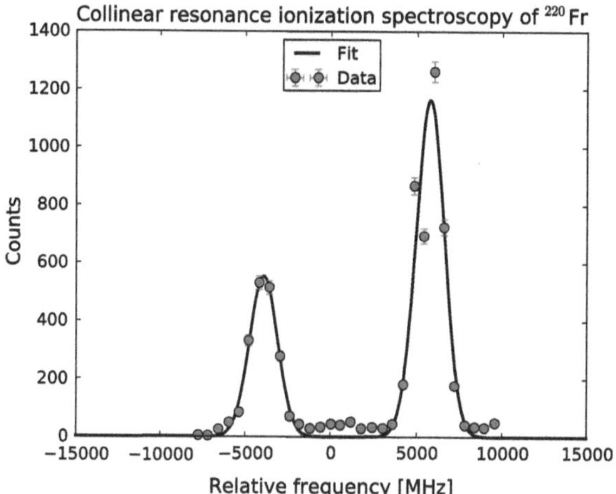

Fig. A.8 Collinear resonance ionization spectroscopy scan of ^{220}Fr

Appendix B
Publications

The results detailed in this thesis have been published by the Author in the following papers:

K.M. Lynch et al., Phys. Rev. X **4**, 011055 (2014)

K.T. Flanagan, K.M. Lynch et al., Phys. Rev. Lett. **111**, 212501 (2013)

K.M. Lynch, T.E. Cocolios, M.M. Rajabali, Hyperfine Interact. **216**, 95 (2013)

K.M. Lynch et al., EPJ Web of Conferences **63**, 01007 (2013)

M.M. Rajabali, K.M. Lynch, T.E. Cocolios et al., Nucl. Instrum. Methods Phys. Res. A **707**, 35 (2013)

K.M. Lynch et al., J. Phys.: Conf. Ser. **381**, 012128 (2012)

The Author also participated on several other experiments during her Ph.D., leading to the publications listed below:

S. Rothe et at., Nature Comms. **4**, 1835 (2013)

T.E. Cocolios et al, Nucl. Instrum. Methods Phys. Res. B **317, Part B**, 565 (2013)

D.A. Fink et al, Nucl. Instrum. Methods Phys. Res. B **317, Part B**, 417 (2013)

B.A. Marsh et al, Nucl. Instrum. Methods Phys. Res. B **317, Part B**, 550 (2013)

T.J. Procter et al., Phys. Rev. C **86**, 034329 (2012)

T.J. Procter et al., J. Phys.: Conf. Ser. **381**, 012070 (2012)

B. Cheal et al., J. Phys.: Conf. Ser. **381**, 012071 (2012)

K.M. Lynch, *Laser Assisted Nuclear Decay Spectroscopy*,
Springer Theses, DOI: 10.1007/978-3-319-07112-1

Zeitfracht Medien GmbH
Ferdinand-Jühlke-Straße 7
99095 Erfurt, Deutschland
produktsicherheit@kolibri360.de